# Dictionary of Electrical, Electronics and Computer Abbreviations

*To*
Sue, Camilla and Estella
*with thanks*

# Dictionary of Electrical, Electronics and Computer Abbreviations

## Phil Brown
### TEng, MIGTechE, MIED
Editor: *Electronics News*
Former Editor: *Whats New in Electronics*

Butterworths

London Boston Durban Singapore Sydney Toronto Wellington

**First published 1985**

R 621.3

© **P. R. Brown 1985**

---

**British Library Cataloguing in Publication Data**

Brown, Philip R.
    Dictionary of electrical, electronics and
    computer abbreviations
    1. Electric engineering–Abbreviations
    I. Title
    621.3′0148        TK9

    ISBN 0-408-01210-2

---

**Library of Congress Cataloging in Publication Data**

Brown, P. R. (Philip Raymond)
    Dictionary of electrical, electronics, and computer
    abbreviations

    Bibliography: p.
    Includes index.
    1. Electric engineering–Abbreviations–Dictionaries.
    2. Electronics–Abbreviations–Dictionaries.
    3. Computers–Abbreviations–Dictionaries.
    I. Title.
    TK9.B76   1985      621.3′0148     84-28514

    ISBN 0-408-01210-2

---

Photoset by Butterworths Litho Preparation Department
Printed and bound in Great Britain by Anchor-Brendon Ltd, Tiptree, Essex

# Preface

Abbreviations can communicate information and provide understanding of complex ideas succinctly and swiftly. Their derivation, however, can be from numerous sources, brought about by the curtailment of principles during general usage or by the 'coining' of labels at the initial or conceptual stage.

Though the principles of abbreviations might appear simple to those who use them, their derivation can be haphazard because there are no specific rules or laws governing the production of acronyms or abbreviations – either they are born or they evolve. For example, the 'dual-in-line' concept has been logically curtailed to DIL, so that items are referred to as 'DIL sockets'; but when the housings or packages employed in the DIL system are used they are not referred to as 'DILP items' (dual-in-line packages) as one might expect but are termed 'DIP items'. Therefore, guesswork when employing or trying to understand abbreviations can lead to misunderstanding in communication.

The material for this book has been drawn from a number of subject areas, including: electrical, electronics, computers, telecommunications, fibre optics, microcomputers/microprocessors, audio, video, information technology, avionics, military, data processing, instrumentation, units, measurement, standards, services, organizations, associations and companies.

The mix has been selected to provide a comprehensive and broad view of 'electronics' and all that is associated with it, and as there is now generally an overlapping of disciplines (e.g. fibre-optic terms in data processing, computers within electronics) this dictionary can assist as an *aide-mémoire* when reading specification sheets and other technical publications, and when keeping up-to-date with institution journals and trade magazines.

Cross-references have been provided in the form of abbreviations within parentheses. For example, the entry for CCD, charge-coupled device, provides three cross-references: charge-transfer device (CTD), metal-oxide semiconductor (MOS), and capacitor (C,6). To broaden further the explanation, if required, referral to entries CTD, MOS and C part 6 (C is a multiple entry with nine parts) is advised. The system of prefixes and units used in the Système International d'Unités has been followed throughout this book. For further clarification, if required, the reader is referred to the entry SI,1.

For the engineer or technician this dictionary will serve as a memory jogger which involves them to use the cross-references to obtain a greater understanding of the subject. Whilst for sales and purchasing personnel, use as a reference companion to keep abreast of commonly-used terms is recommended as the most rewarding and time-saving employment of this book.

If further information is required the following publications are recommended:

*Dictionary of Audio, Radio and Video* by R. S. Roberts
*Dictionary of Data Processing* (2nd ed) by Jeff Maynard
*Dictionary of Electronics* by S. W. Amos
*Dictionary of Electrical Engineering* (2nd ed) by K. G. Jackson, revised by R. Feinberg
*Dictionary of Telecommunications* by S. J. Aires
*Electronics Engineer's Reference Book* (5th ed) by F. Mazda

P.R.B.<br>Shooters Hill

# A

**a**  Atto. A decimal prefix commonly used in association with a base unit in Le Système International d'Unités (SI) system of units, indicating a multiplication factor of $10^{-18}$ of that unit. For example, one attosecond, as, is equal to 0.000 000 000 000 000 001 of a second.

**A**  Ampere. The SI unit of electric current, named after the French scientist André-Marie Ampere (1775–1836). It is the electric current (I) which, when flowing in two infinitely-long parallel conductors, with their centres 1 metre (m) apart in a vacuum, causes each conductor to experience a force of 2 × $10^{-7}$ newtons per metre length of conductor.

**Å**  Ångstrom. A unit of length equal to $10^{-10}$ m. It is often used to express the wavelengths of very-high-frequency signals such as light.

**AA**  One of the letter designations given to the most commonly-used sizes of 1.5-volt (V,1) batteries. Originally of North American nomenclature, this type of letter description is gaining increasing popularity over the International Electrotechnical Commission (IEC) letter/number system as it is able to simplify the selection of batteries. *See* battery-size cross-reference guide, Appendix.

**AAA**  Another of the letter designations given to the most commonly-used sizes of 1.5-volt (V,1) batteries. *See* AA above.

**AB**  (1) AB pack. A term used in the USA to describe a complete portable power source for electron tubes, comprising the A battery supplying the heater power and the B battery supplying the anode circuit power.

(2) Automated Bibliography. A history and description of a number of books, a list of documents pertaining to a given author or an annotated catalogue of documents held in a computer file.

**ABA**  American Bankers' Association. An association that promotes aids for its members. An example is ABA coding; numbers that assist computerized cheque clearance.

1

**ABC** Automatic Brightness Control. A circuit used to adjust automatically the average luminance (brightness) level of a display to compensate for changes in ambient light level.

**ABCA** America, Britain, Canada and Australia Standard. Representatives of the four governments participate in the meetings of this body to agree on standards, especially those applying to military suppliers and industry.

**ABEND** ABnormal END. An irregular or freakish termination to a job or computer operation following either an error condition or an intervention by the operator.

**ABL** Atlas BASIC Language. A computer language designed for use on an Atlas computer.

**ABS** Acrynolite Butadine Styrene. A mouldable plastics compound that provides a rigid, tough and scratch-proof material.

**AC** (1) Alternating Current. An electric current (I) that has its direction in a circuit (CCT) reversed with a frequency (*see* Hz) that is independent of circuit constants. Alternating currents are normally represented by a sine wave, as shown in *Figure A.1*.

(2) ACcumulator. *See* ACC.

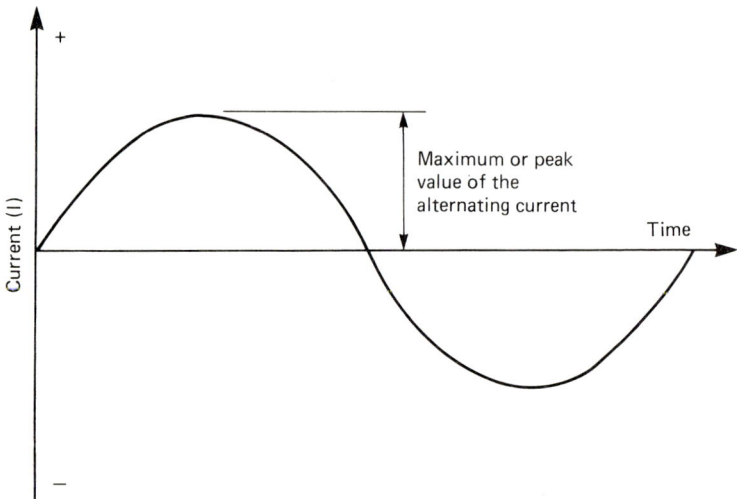

*Figure A.1*  One complete cycle of an alternating current (AC,1)

**ACARD** Advisory Council for Applied Research and Development. A body in the UK that provides the government with advice concerning research and development (R & D) policies, as well as publishing reports.

2

**ACC** ACCumulator. (a) A secondary cell having a reversible chemical action and charged by passing a current (I) through it. Direction and rate of chemical action are determined by the value of the external voltage (V). (b) A register of the arithmetic/logic unit (ALU) of a central processor unit (CPU) used as intermediate storage during the formation of algebraic sums, or for other intermediate logical and arithmetic operations.

**ACE** Automatic Computing Engine. A name given to one of the earliest computers.

**ACIA** Asynchronous Communications Interface Adaptor. A device that provides the data formatting and control to interface serial asynchronous data communications information to bus organized systems (i.e. between a microprocessor and a modulator/demodulator (MODEM)).

**ACK** ACKnowledge. A character code indicating a positive acknowledgement that a message has been received correctly, or that a receiver is ready to receive data.

**ACM** Association for Computing Machinery. A professional and technical society whose activities, publications and conferences are designed to aid computing advancement. The aid is with specific regard to machinery and system design, language and program development, plus other related activities.

**ACOMPLIS** A COMPuterized London Information Service. Held on a computer, this is an information service run by the Greater London Council's Library Department.

**ACRE** Automatic Call-Recording Equipment. An invention of British Telecom (BT) which dispenses with the need for operator written 'tickets' detailing calls they handle. ACRE records the details automatically on magnetic tape for subsequent processing by BT's computer-based telephone billing system.

**ACTP** Advanced Computer Technology Project. Funded by government in the UK, this is a project for research into and development of advanced computer techniques for industry.

**ACU** Automatic Calling Unit. A dialling device that permits a business machine to dial calls automatically over a communications network.

**A/D** Analogue-to-Digital. (a) The process of converting between an analogue and a digital signal. (b) A circuit used to convert information into digital form from an analogue input. Such a circuit is used in digital voltmeters and similar devices. *See* ADC, analogue-to-digital converter.

**ADAPSO** Association of DAta Processing Service Organisations. A data processing (DP,1) organization that also includes

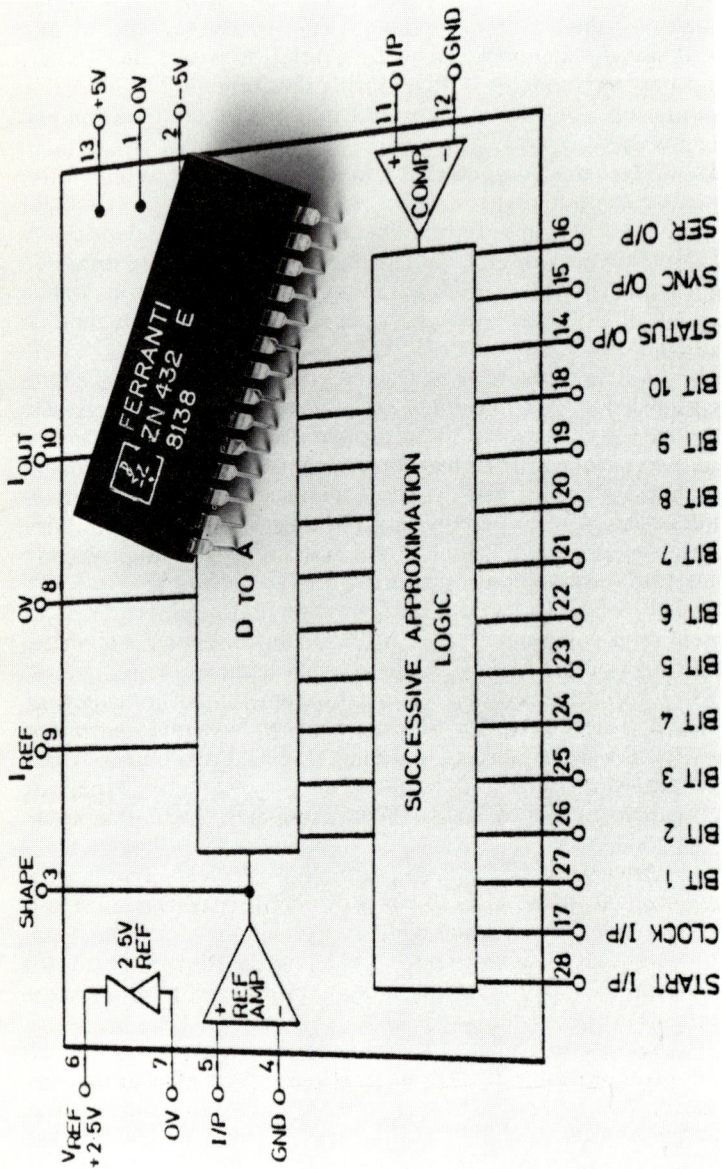

Figure A.2 Ten-bit analogue-to-digital converter(ADC) in a 28-pin package (*courtesy* Ferranti Electronics Ltd.)

4

a software development and marketing group.

**ADC(A/DC)**   Analogue-to-Digital Converter. A device or circuit that changes analogue input voltages to their equivalent digital (binary or binary-coded decimal) values for acceptance by the memory of a digital processor. An 8-bit ADC generates an 8-bit word; a 12-bit ADC generates a 12-bit word. The encoding becomes more precise as more bits are used.

ADCs fall into three main groups, (a) integrating converters, (b) director flash converters and (c) feedback converters with internal self-checking digital-to-analogue converters (DAC) as shown in *Figure A.2.*

**ADCCP**   Advanced Data Communications Control Procedures. A version of synchronous data link control (SDLC) standards produced by the American National Standards Institute (ANSI).

**ADCON**   ADdress CONstant. A value or expression used to calculate the real or virtual storage (VS) address of a memory.

**ADF**   Automatic Direction Finder. Typically an ADF consists of two crossed ferrite-core loops projecting 25 mm from the skin of an aircraft with an electronic goniometer generating a direct-reading display of bearing to the selected transmitter. For preference they are tuned to inland non-directional beacons (NDB), which exhibit a sharp reduction in signal strength when directly overhead.

**ADLC**   Advanced Data Link Controller. Data that is transmitted and received in a synchronous serial form in a data communications system is converted into parallel form, analysed and stored by this equipment, so that data link management can be accomplished.

**ADP**   (1) Automatic Data Processing. (a) Equipment such as electronic accounting machines and electronic data processing (DP,1) units or systems. (b) Data processing (DP,1) performed by a system of electronic or electrical machines interconnected and interfacing so as to reduce to a minimum the need for human assistance or intervention.

(2) Automatic Die Positioner. An item of production equipment that automatically selects and aligns good dies on a silicon wafer. The dies are collected by an automatic mechanical probe and bonded to the substrate of a chip (CHIP) package.

**ADPE**   Automatic Data Processing Equipment. Machinery employed to perform automatic data processing (ADP,1(b)) functions.

**ADPS**   Automatic Data Processing System. A system employed to perform automatic data processing (ADP,1(b)) functions.

**ADRS**   Automatic Document Request Service. Provided by the

British Library Automated Information Service (BLAISE), this service permits subscribers at an on-line terminal to ask for the loan of, or copies of, documents from the British Library Lending Division.

**ADU** Automatic Dialling Unit. A device capable of automatically generating telephone dialling digitizer signals.

**AECMA** Association Européene des Constructeurs de Material Aerospatial. Concerned with aerospace matters, this association prepares draft standards that are subject to normal European Committee for Standardisation (CEN) procedures before becoming adopted as European Standards.

**AF** Audio Frequency. Any frequency that can be detected by a normal human ear. This audile range extends from about 20 to 20 000 Hz. In a communication system intelligible speech can be obtained if a frequency range from about 300 to 3400 Hz is reproduced; any frequency in this range is called a voice frequency (VF).

**AFC** Automatic Frequency Control. A device that automatically maintains the frequency of an alternating voltage within specified limits. Error-operated, the device normally consists of two parts, a frequency discriminator that compares actual with desired frequencies, and a reactor controlled by the discriminator in such a way as to correct any frequency error. *See Figure A.3.*

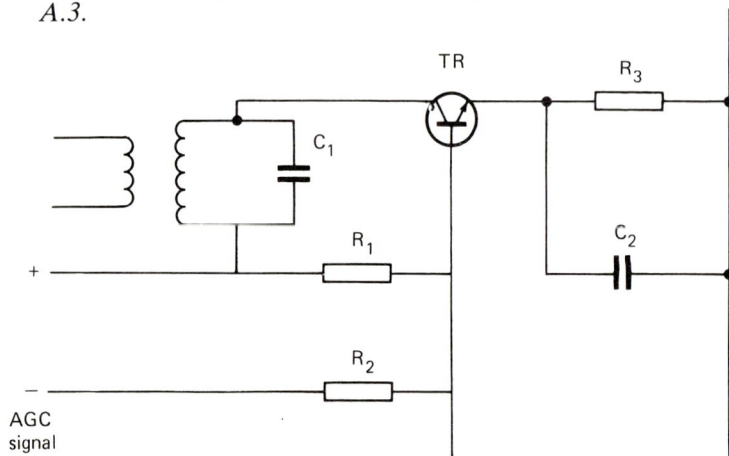

*Figure A.3* Automatic frequency control (AFC) undertaken by a typical intermediate frequency (IF,1) stage

**AFIPS** American Federation of Information Processing Studies. An association of various USA data processing (DP,1) groups, institutes, societies and organizations. Formerly 'AFID'.

**AFNOR** Association Français de NORmalisation. A standards authority in France that is a member of the International Standards Organisation (ISO).

**AFSK** Audio Frequency-Shift Keying. A system employed in telecommunications that uses audio frequency (AF) tones to modulate a carrier wave in order to convey a digital signal.

**Ag** Argentum. Silver. A white precious metal that is an excellent electrical conductor. Atomic number, 47; atomic weight 107.880.

**AGC** Automatic Gain Control. (a) A device that maintains the output volume of a radio receiver at a constant level, despite fluctuations in the input signal. An AGC is usually applied to amplifiers. (b) The process of maintaining the output volume of a radio receiver at a constant level.

**AIM** Avalanche-Induced Migration. An irreversible process employed in the programming of a programmable read-only memory (PROM), based on aluminium migration through silicon to short circuit a diffused junction. *See Figure A.4.*

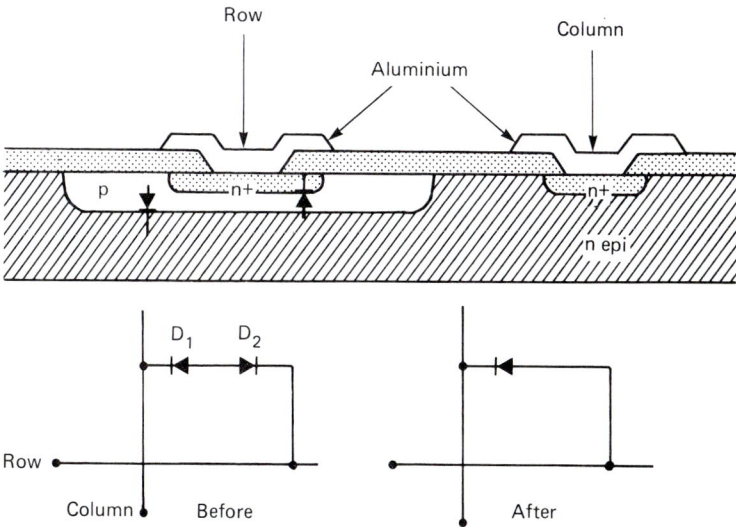

*Figure A.4* The basic structure of an avalanche-induced migration (AIM) cell

**AIO** Action Information Organization. An operations room on a warship that takes inputs from aircraft communications, radar, sonar, computerized gun and missile controls and internal communications. Such information is displayed on consoles so that necessary human or computer intervention can be made, for instance to take evasive action when under attack.

**AKWIC** Author and Key Word In Context. An index that lists available computer programs or documents tabulated in alphabetical order of the most informative or significant word in the title of the information, along with the writer or programmer of the particular item of data.

**ALGOL** (1) ALGOrithmic Language. A computer language by which numerical procedures may be precisely presented to a computer.

(2) ALGebraic Orientated Language. The international procedural computer programming language used especially for scientific applications.

**ALOFT** Airborne Light Optical Fibre Technology. A system used for the economic analysis of fibre optics technology in the American A-7 aircraft.

**ALPHAMERIC** A contraction of 'alphanumeric'.

**ALS TTL** Advanced Lower power Schottkey Transistor-Transistor Logic. A recent addition to the transistor-transistor logic (TTL) range of devices, the ALSTTL employs improved integrated circuit (IC) construction techniques, permitting smaller geometry on the integrated circuit slice, which provides higher speeds for no power increase. It is available with a propagation delay of 4.5 nanoseconds and a power consumption of 1.5 milliwatt.

**ALU** Arithmetic and Logical Unit. The register or portion of hardware of a computer or processor in which binary arithmetic and logical functions are performed, as shown in *Figure A.5.*

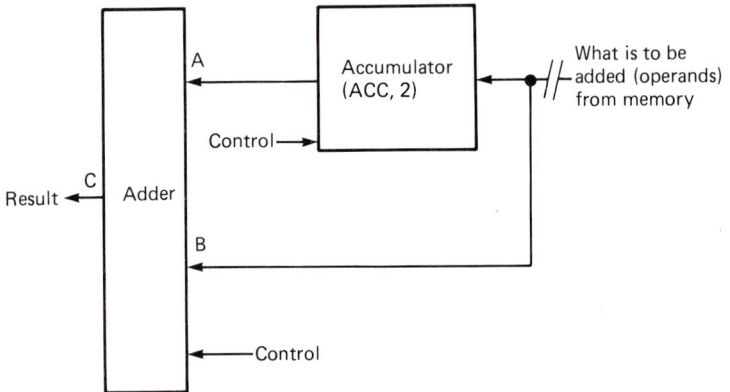

*Figure A.5* One particular type of arithmetic and logical unit (ALU)

**AM** Amplitude Modulation. (a) A modulation in which the peak values, positive and negative, of an alternating current (AC) or wave are varied above and below their unmodulated values by

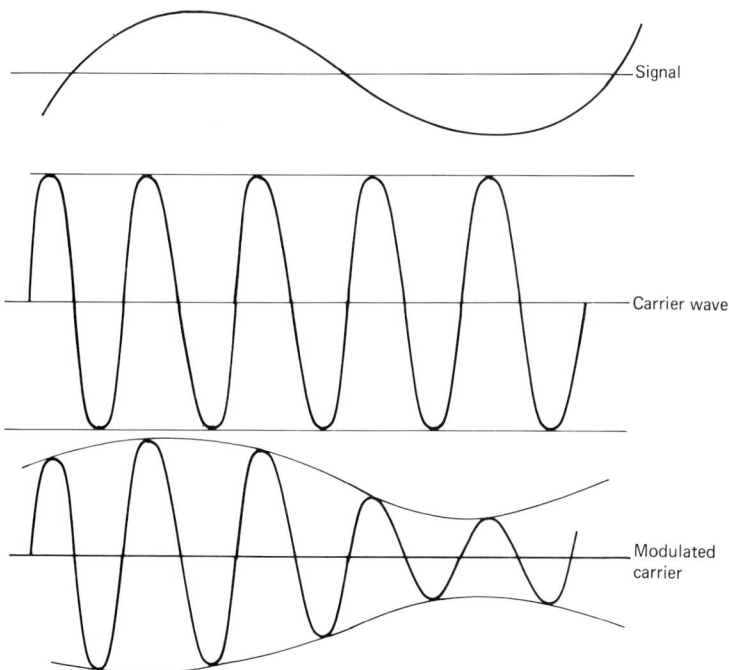

*Figure A.6* Amplitude modulation (AM,(a))

an amount proportional to the peak value of the signal wave and at the frequency of the modulating signal, as shown in *Figure A.6*. (*b*) Also refers to a radio receiver or part of a radio receiver that detects amplitude-modulated signals only.

**AM1** Air Mass ONE. A term used in connection with solar cells. When the Sun is directly overhead, the length of its path through the atmosphere is at its shortest; thus the optical air mass is considered unity and the radiation is described as AM1.

**AMACUS** Automated Microfilm Aperture Card Update System. A system employing aperture cards, consisting of an 80-column punched card with a $35 \times 45$ mm frame of microfilm inserted, as its main form of storing information.

**AMBIT** Algebraic Manipulation BIT. A particular computer programming language used for algebraic symbol manipulation.

**AM/FM** Amplitude Modulated/Frequency Modulated. Usually refers to a radio receiver that detects both amplitude modulated (AM) and frequency modulated (FM) signals.

**AMO** Air Mass ZERO. Also known as the 'solar constant', AMO is generally used in relation to solar energy cells. It can be described as the radiant power per unit area perpendicular to the direction of the Sun outside the Earth's atmosphere, but at the mean Earth-Sun distance. The radiation intensity is referred to as the AMO radiation.

**AMP** (1) AMPlifier. A circuit (CCT) device or stand alone unit that is capable of increasing the magnitude of a given electrical input signal employing energy taken from an external source (i.e. it introduces gain). There are many types of amplifier and the particular nomenclature is dependent upon their application, construction and whether they are operated by an alternating current (AC, 1) or a direct current (DC, 1). But basically there are two main groups: linear amplifiers (where the output (O, 2) signal is a linear function of the input signal) and non-linear amplifiers. *See Figure A.7.*

(2) A contraction of the word Ampere. *See A.*

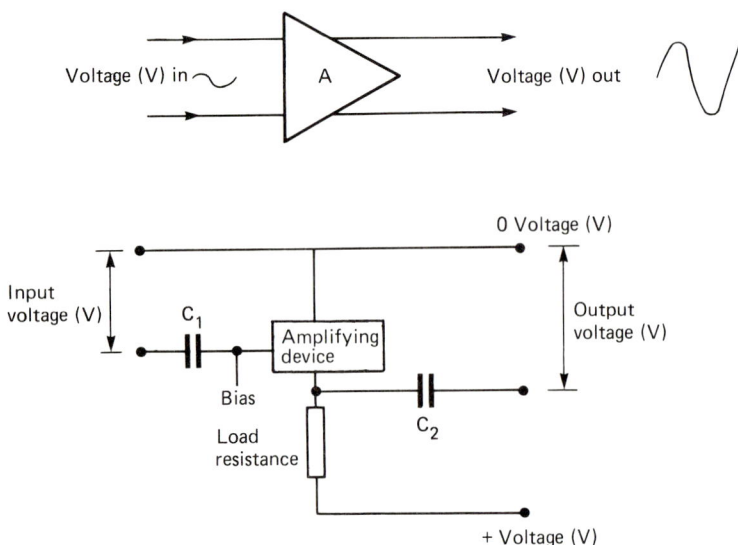

*Figure A.7* Schematic diagram of an amplifier (AMP) and a typical amplifier circuit (CCT)

**AMSAT** The Radio AMateur SATellite Corporation. A corporation formed in the USA in 1969 to co-ordinate a range of American and international amateur space projects. *See* OSCAR.

**AND** A logic function, operation, circuit or gate. (a) A logical function or operation defined by a specific rule that if two statements, A and B, are true (1), then C is also true (1); if not, then C is false (0). Truth is usually expressed by the value 1, while 0 is used to indicate a false state, as shown in *Figure A.8*. (b) A physical circuit that performs the AND function. (c) A gate that performs the AND function.

**ANL** Automatic Noise Limiter. A circuit employed in radio transmitter/receivers, now extensively used in citizen band (CB) transceivers to reduce noise or interference such as that caused by motor-car ignition systems.

**ANSI** American National Standards Institute. An organization that establishes standards and organizes committees of computer users and manufacturers. It develops and publishes industrial standards in the fields of computing and information handling that are accepted world wide. Formerly USASI.

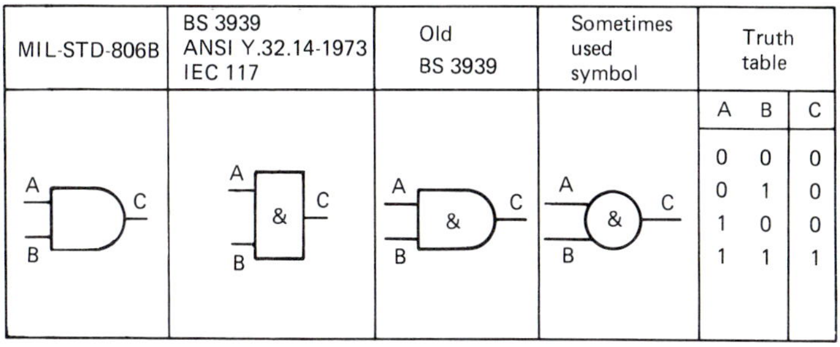

| MIL-STD-806B | BS 3939 ANSI Y.32.14-1973 IEC 117 | Old BS 3939 | Sometimes used symbol | Truth table | | |
|---|---|---|---|---|---|---|
| | | | | A | B | C |
| | | | | 0 | 0 | 0 |
| | | | | 0 | 1 | 0 |
| | | | | 1 | 0 | 0 |
| | | | | 1 | 1 | 1 |

*Figure A.8* The AND gate. In practice the MIL-STD-806B symbol is the one most likely to be met, although the other logic symbols may be encountered.

**A-O** Acousto-Optic. A term that refers to the interaction of optical and acoustic waves.

**APD** Avalanche PhotoDiode. A photo-detecting diode sensitive to light energy. It increases its electrical conductivity by increasing the number of electrons in its conduction band through absorbing photons of energy, the interaction of electrons, and an applied bias voltage.

**APL** Algorithmic/Procedural Language. A programming language with an extensive set of operations and data structures that are used to implement what is considered by many to be the most flexible, powerful and concise language in existence. It was invented by Kenneth Iverson and is also known as 'A Programming Language'.

**APT** Automatically Programmed Tools. A system for the computer-assisted programming of numerically controlled (N/C) machine tools, draughting machines and similar equipment.

**AQL** Acceptable Quality Level. Batches of components or parts are inspected and laboratory tested by many manufacturers to a specific percentage value. If more than that percentage fail to meet the AQL, the whole batch is rejected.

**ARPA** Advanced Research Projects Agency. An agency of the US Department of Defense, which supports its own resource-sharing computer network (ARPANET).

**ARPANET** Advanced Research Projects Agency NETwork. A large packet-switching network that employs exchanges linked by high-speed data-transmission facilities, developed by the US Department of Defense.

**ARQ** Automatic ReQuest for repetition. An automatic system that provides error correction by making use of a code and a construction such that any false signal initiates a repetition of the transmission of the character incorrectly received. Also known as 'automatic error correction'.

**ARTEMIS** Automatic Retrieval of Test through European Multipurpose Information Services. Currently under evaluation by the European Commission, ARTEMIS is planned to be a system that can deliver documents, supplying articles of text on demand over a facsimile (FAX) or digitized text-transmission system.

**ARU** Audio Response Unit. A device that can connect a computer system to a telephone and thereby provide a voice response to any enquiries received.

**ASA** American Standards Association. A US association responsible for the establishment of standards; formerly known as the United States of America Standards Association.

**ASC** Automatic Sequence Control. A management feature found in computer programs.

**ASCC** Automatic Sequence Controlled Calculator. An early computer constructed by Howard Aiken at Harvard University in 1937.

**ASCII** American Standard Code for Information Interchange. Usually pronounced *ask-ee*. This is an 8-bit information code for data transfer used with most computer and data terminals. It has been adopted by the American Standards Association (ASA) to achieve compatibility between data devices and other pieces of equipment.

**ASP** Attached Support Processor. A term used to describe multiple computers, usually two, connected via channel-to-channel adaptors, employed to increase efficiency in processing many short duration tasks.

**ASR** Automatic Send-Receive set. A combination teletypewriter (TTY), transmitter and receiver having the capability to transmit from either a keyboard or paper tape.

**ASW** Anti-Submarine Warfare. A typical acoustic processing and display advanced airborne anti-submarine defence system comprising receivers, digital data processors, operator controls and the displays on which information about a submarine is presented.

**ATE** Automatic Test Equipment. Equipment or systems that can test devices automatically. For complex devices, especially semiconductor devices, the systems are usually programmable. They may be purpose-built or constructed from units employing the general purpose interface bus (GPIB) standard. Operators are needed only to connect the unit under test (UUT) to the system.

**ATLAS** Automatic Tabulating, Listing And Sorting system. A package of computer software that is used for tabulating, listing and sorting.

**ATR** Anti-Transmit-Receive. A switch that automatically decouples a RADAR transmitter from its antenna during a receiving period; extra protection is thus provided from return echoes received by the antenna. *See* (TR) Transmit-Receive.

**ATS** Application Technology Satellite. Used in particular for medical communications, this is the designation given to a series of American communications satellites that provide both audio and video channels.

**ATU** Antenna Tuning Unit. A specific circuit that adjusts and varies the combined effects of capacitance (C, 3), resistance and inductance of an antenna to match that of transmitters and receivers.

**AUTOABSTRACT** AUTOmatic ABSTRACT. Significant words from the title and sometimes the body of a document that describe its contents are selected and arranged in a meaningful pattern. The arrangement and selection by computer is carried out according to a specific programmed standard. Although not looking like an abstract prepared by humans, the autoabstract normally provides a good indication of what the document is about.

**AUTOPOLL** AUTOmatic POLLing. Without interrupting the central processing unit (CPU), this hardware feature is able to process a list specifying the order in which a number of computer terminals are polled. It also handles negative responses.

**AUTOVERIFIER** AUTOmatic VERIFIER. Equipment employed to read information punched on a card and verify that information.

**AVC** Automatic Volume Control. A term that is still in use but should be deprecated in favour of 'automatic gain control'. (*See* AGC(a)).

*Figure A.9* An AZERTY keyboard (*courtesy* Alphameric Keyboards Ltd.)

14

**AVPO**   Axial Vapour-Phase Oxidation process. A process used for making graded-index (GI) optical fibres (OF) in which the glass preform is grown radially rather than longitudinally as in other processes.

**AWG**   American Wire Gauge. A system of gauging or sizing wires that provides a particular number as a designation for a specific diameter of wire and is frequently employed in the USA. It is also known as Brown & Sharpe gauge (B & S).

**AZERTY**   A French language computer terminal or typewriter keyboard which has the keys arranged such that the top row below the numerals begins with the letters, A, Z, E, R, T and Y. *See Figure A.9.* It normally contains some additional letters peculiar to the French language, as well as instructional keys notated in French. *See* QWERTY and QWERTZ.

## B

**B**   Battery. Nowadays usually the B part of the AB pack that supplies power to the anode circuit of an electron tube. In the early days of radio the B battery was a multi-cell Leclanché A battery which was a lead-acid accumulator (ACC(a)). *See* AB,1.

**BAEC**   British Amateur Electronics Club. A national amateur electronics club in the UK that helps all who are interested in electronics. Started in 1966, it has many members from beginners to experts in all parts of the British Isles and overseas.

**BAL**   Basic Assembler Language. A non-complex computer program that translates an assembly language into object code, having no macros or conditional assembly instructions.

**BALUN**   BALanced to UNbalanced transformer. Typically this type of transformer is employed to couple a balanced antenna to an unbalanced transmission line.

**BAM**   Basic Access Method. Each input/output (I/O) statement in this method of computer access initiates the performance of a machine input/output operation.

**B & S**   Brown & Sharpe gauge. A system of gauging or sizing of wires that provides a particular number as a designation for a specific diameter of wire and is frequently employed in the USA. Also known as 'American wire gauge' (AWG).

**BASIC**   Beginners All-purpose Symbolic Instruction Code. A popular computer language characterized by its ease of use and the ease with which it can be learned. It is the standard language used with microcomputers and is well-suited to computer time-sharing.

**BASIS** British Airways Staff Information System. A videotex system employed by British Airways for conveying information.

**BAT** BATtery. A source of direct current (DC,1) or voltage (V), producing electrical energy by chemical means, it consists of one or more electrolytic cells connected together and used as a single unit.

**BATS** Basic Additional Teleprocessing Support. A system used by some computer systems in the late 1960s to provide telecommunication processing support.

**BAUD** BAUDot. A unit of measure of data signalling rate or data flow. It is the instantaneous bit speed with which a device or channel (CHNL(b)) transmits a character in serial transmission. Although strictly speaking it stands for 'signal speed', for practical purposes it is generally considered as interchangeable with bits per second (BPS). Emil Baudot produced the first widely-used teleprinter in the nineteenth century, as well as the Baudot code for use on telegraphic circuits.

**BAVIP** British Association of Viewdata Information Providers. A body comprised of individuals or organizations that provides information or data for viewdata systems.

**BBC** British Broadcasting Corporation. The independent non-commercial radio and television authority in the UK, that is funded by the legally enforced licensing of television receivers.

**BBD** Bucket-Brigade Device. A device consisting of capacitors linked by switches. The switches in practice are bipolar or metal-oxide silicon field-effect transistors (MOSFETs) as shown in *Figure B.1*. These circuits are now usually made as integrated circuits (ICs). Clocked or timed pulses are applied to close the switches, a two-phase system (A and B) being used. As each switch is closed charge is transferred from one capacitor to another.

**B-BOX** (a) A register that contains a quantity to be used under direction of the control section of computer hardware. (b) A device that permits automatic modification of an instruction address without permanently altering the instruction in the memory. (c) A register to an arbitrary integer, usually one which is added (or subtracted) upon the execution of each machine instruction.

**BCC** Block Check Characters. Check bits used in line control procedures, which make up the later part of a block of transmitted information. They are generated by a checking process applied to the information part of the transmission.

**BCCD** Bulk-channel Charge-Coupled Device. A charge-coupled device (CCD) that has formed, by epitaxy or ion implantation, a thin layer on the surface of the substrate of an opposite

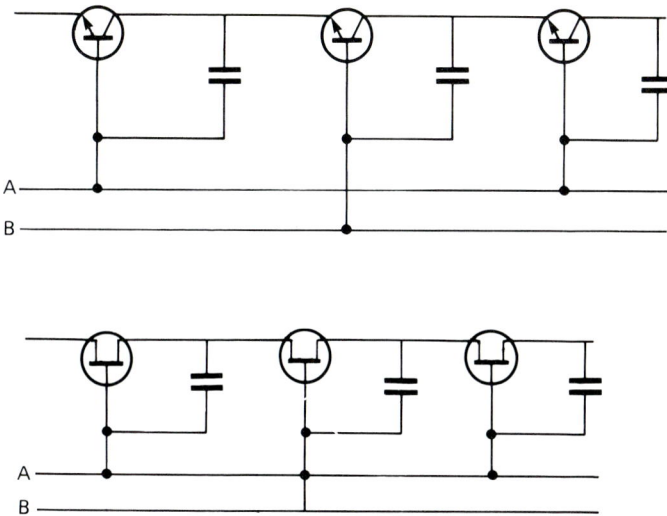

Figure B.1 Bucket-brigade devices (BBD): top employing bipolar transistors; bottom employing metal-oxide silicon field-effect transistors (MOSFET)

conductivity type. It thus confines the flow of charges to a channel (CHNL(h)) lying beneath the surface of the semiconductor. Also known as a 'buried-channel charge-coupled device'. *Figure B.2.*

**BCD** Binary Coded Decimal. A numerical representation in which digits are represented by binary numerals (1 or 0), and the individual figures of a decimal number are separately coded into binary form. The most common code uses a presentation

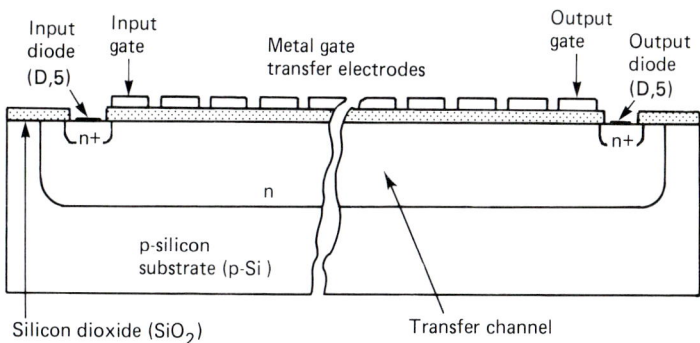

Figure B.2 A cross-sectional view of a bulk-(buried) channel charge-coupled device (BCCD)

17

where 8-4-2-1 binary code groups are used, as shown in *Figure B.3*. In this code twenty three is represented as 0010 0011; in pure binary notation it would be 10111.

**BCO** Binary Coded Octal. This is a system where binary numbers are employed to represent the octal digits of an octal number.

**BCPL** Basic Combined Programming Language. Developed to write a computer program that translates from a high level into object code for the combined programming language (CPL), this system's implementation language is available on a wide range of machines.

| Binary | Decimal | Binary | Decimal |
|--------|---------|--------|---------|
| 0000 | 0 | 0101 | 5 |
| 0001 | 1 | 0110 | 6 |
| 0010 | 2 | 0111 | 7 |
| 0011 | 3 | 1000 | 8 |
| 0100 | 4 | 1001 | 9 |

*Figure B.3* Binary-coded decimal (BCD) numerical representation

**BCS** Superconductivity theory. The BCS theory in which pairs of electrons (Cooper pairs) combine in the presence of other electrons. Interactions between electrons result in the pairing and vibrations of the crystal lattice providing a material with a highly-ordered state. The material then exhibits little resistance to electron movement with no dissipation of energy.

**BDAM** Basic Data Access Method. Providing a non-queueing facility, BDAM is a method employed by some computers of gaining access to data held in a backing store. Although not a primary store the backing store has in fact a larger capacity but a slower access time.

**BDOS** Basic Disk Operating System. An area in the memory map (MAP,2) of a microcomputer using a control program for microcomputers (CP/M) disk operating system (DOS), having routines used by the system to access its system disk drives. It thus provides for file management.

**BDV** BreakDown Voltage. The voltage under specified conditions at which an equipment breakdown occurs. Compare BV which applies only to semiconductor devices.

**BEEF** Business and Engineering Enriched Formula translator. Employed in business and engineering applications, this is an enhanced version of formula translator (FORTRAN) computer language.

**BEL**   BELl character. One of the range of characters that form a specific code for any particular system (character set) that will initiate the sounding of a bell on a terminal device.

**BER**   Bit Error Rate. This is usually expressed as the number of erroneous bits per million introduced in the passage of single or parallel streams of binary digits (BITs) in a transmission path.

**BEX**   Broadband EXchange. In the USA a public switched communication system of the Western Union Company. It features various bandwidth full duplex (FDX) connections.

**BFL**   Back Focal Length. The distance measured from the vertex of the back surface of a lens to its rear focal point.

**BFO**   Beat Frequency Oscillator. A circuit generating an alternating current (AC, 1) with an adjustable frequency. Its output can be mixed with that of a final intermediate-frequency amplifier (AMP). It produces an audible frequency beat when the receiver is tuned to an unmodulated signal.

**BH**   Busy Hour. An uninterrupted period of one hour during which the average intensity of traffic of a telephone exchange or group of telephone circuits is at a maximum.

**BIC**   Bipolar Integrated Circuit. A particular type of monolithic integrated circuit (IC,1), that has its basic structure of bipolar transistors.

**BIDAP**   BIbliographic DAta Processing program. A package of computer software designed to process information relating to books, articles or reports. Normally the data covered is of authorship, title and publication details.

**billi**   BILLIon. A prefix to a measure value to indicate a billion units, e.g. billipounds. In the UK a billion has traditionally been a million millions ($10^{12}$), whereas in the USA and elsewhere it stands for a thousand millions ($10^9$).

**BIM**   Beginning of Information Marker. A reflective spot about 3 metres from the physical end of a magnetic tape (MAGTAPE). It is sensed by a photoelectric device and indicates the point on the tape at which recording may begin.

**BIMOS**   BIpolar and MOS techniques. Techniques that combine on a single CHIP both bipolar and metal oxide semiconductor (MOS) devices. These high-density devices offer near-ideal operational-amplifier parameters and enable digital (D,4) and linear functions to exist on the same integrated circuit (IC).

**BIN**   Basic Identification Number. The BIN code is a three-colour-band system on the contact body of avionic rectangular connectors to identify each contact size.

**BINAC**   BInary Automatic Computer. A computer built by the Eckett-Mauchly Corporation in the USA in the late 1960s.

**BIOD**   Bell Integrated Optical Device. An integrated optical circuit (IOC) having both active and passive elements, which can be used as logic and control elements in various other devices (e.g. logic gates) depending on how they are interconnected during fabrication.

**BIOS** Basic Input Output System. A group of routines comprising an area in the memory map of a microcomputer using a control program for microcomputers (CP/M) disk operating system (DOS), that is implemented when a user first gets the system up and running. The area is then called up by the operating system when it needs any input/output (I/O) routines.

**BISAM** Basic Indexed Sequential Access Method. A method employed by some computers to gain access to information held in a backing store, which is not a primary memory store. It is in fact a store that has a larger capacity but a slower access time.

**BISYNC** BInary SYNChronous communications. In a data communications system, BISYNC is the use of a set of control characters and sequences for the synchronized transmissions of binary coded data between stations. It is a low-level half-duplex (HD) communications protocol that allows any pattern of binary digits (BITs), including those normally used for control, to be transmitted in a block. Also called BSC.

| Equivalent number (decimal) | Column values (decimal) | | | | |
|---|---|---|---|---|---|
| | 8 | 4 | 2 | 1 | |
| 0 | 0 | 0 | 0 | 0 | |
| 1 | 0 | 0 | 0 | 1 | |
| 2 | 0 | 0 | 1 | 0 | Binary digits (BITs) |
| 3 | 0 | 0 | 1 | 1 | |
| 4 | 0 | 1 | 0 | 0 | |
| 5 | 0 | 1 | 0 | 1 | |
| 6 | 0 | 1 | 1 | 0 | |
| 7 | 0 | 1 | 1 | 1 | |
| 8 | 1 | 0 | 0 | 0 | |
| 9 | 1 | 0 | 0 | 1 | |

*Figure B.4* Binary digit (BIT) numerical representation

**BIT** BInary digiT. The smallest amount of information, consisting of two alternatives, either 0 or 1. It is one character of a digital word, and the position of the bit in the word usually determines its significance. For example, *Figure B.4* shows the column values compared with decimal numbers. The numbers can be obtained by progressing right to left across the columns and adding values when they appear; for example, 9 is 1+0+0+8, right to left.

**BJT** Base Junction Transistor. The centre semiconductor region of a double junction transistor, in which the output current is controlled by the input current.

**BLAISE** British Library Automated Information ServicE. A computer service that contains the authorship, title and publication details of all books published in the UK since 1968. Special software enables the information to be retrieved and edited to produce local catalogues.

**BLAST** Boolean Logic And State Transfer. Modules of design software that permit a designer to define pin neuronics and functions, and to write Boolean equations or state transfer equations which are compiled into a standard integrated fuse logic (IFL) device programming table.

**BLO** BLOcking signal. A signal transmitted to the distant end of a communication circuit to indicate that it has been reserved for use or has been rendered unavailable for use.

**BNC** Bayonet Nut Coupling. A type of connector used for coaxial cable (COAX) radio frequency (RF) interconnection. It employs a two-prong bayonet with a spring coupling unit. BNC connectors are extensively employed within test equipment as they provide a rapid connection/disconnection.

**BNF** Backus Normal Form. A formal computer language structure for the syntax phasing of other computer languages. Also known as 'Backus-Naur form'.

**BNPF** Beginning-Negative-Positive-Finish. An older type of data-encoding output format for programmable read-only memory (PROM) programmers. B represents beginning, N is negative (1), P is positive (O) and F is finish. For example, 101010 is represented by BNPNPNPF.

**BOD** Break Over Diode. A two terminal thyristor that is switched to the low impedance state by exceeding the forward breakover voltage. It is an undirectional device whose forward characteristic is similar to a reverse-biased avalanche diode.

**BOP** Bit Oriented Procotol. Usually refers to messages (i.e. BOP messages) used in telecommunications, that are transmitted in frames, all messages adhering to one standard frame format.

**BOS** Basic Operating System. Designed to run on small IBM computers, this software, held resident on a disk, is associated with the execution of programs and the co-ordination of a computer system.

**BOT** Beginning Of Tape. The point to which a cartridge tape is rewound, as opposed to a beginning of information marker (BIM).

**BPAM** Basic Partitioned Access Method. A method employed by some computers to gain access to information held in a backing store, which is not a primary memory store, but a store that has a larger capacity but a lower access time.

**BPI** Bits Per Inch. An expression of the density of data on a tape or disk memory-storage medium; the number of binary digits (BITs) over a specific length.

**BPS** (1) Bits Per Second. A unit of measure of data flow. It is the instantaneous binary digit (BIT) speed with which a device or channel transmits a character in serial transmission. For practical purposes the term is generally considered to be interchangeable with BAUD.

(2) Basic Programming Support. BPS card or tape systems are simple and small operating systems (OS) available for mainline computers.

**BRA** British Robotic Association. An organization of all who are interested in industrial robots, including suppliers, potential users, government agencies and trade unions. It was founded in 1977 and is UK based.

**BROWSER** BRowsing On-line With SElective Retrieval. A computer system data base able to offer ordinary English language searching of the documents that it holds.

**BRS** Bibliographic Retrieval Services. An American service offering access to over 20 of the most popular computer data files with information relating to books, articles or reports. Normally the information covered is of authorship, title and publication details.

**BS** (1) Back Space. One of the range of characters that form a specific code for any particular system (character set) that will cause the automatic backspacing of an input or output device.

(2) British Standard. A prefix normally used in conjunction with a number and a date to denote a particular standard laid down and published by the British Standard Institution (BSI). For example, BS1991: Part 6: 1975 refers to letters, signs and abbreviations for electrical science and engineering.

**BS 9000** The British Standards 9000 System. A system of quality requirements and fundamental performance characteristics prepared for families of components such as capacitors, connectors, relays, resistors and switches. Various levels are established by agreement between manufacturers, users, and National Supervisory Inspectorate (NSI) officers at the British Standards Institution (BSI). Approval tests can be authorized on behalf of the BSI by NSI officers at a local level.

**BSAM** Basic Sequential Access Method. A computer system with a large continuously updated file of information, abstracts or references concerning a particular subject (i.e. a data base).

**BSC** Binary Synchronous Communications. *See* BISYNC.

**BSI** British Standards Institution. An authority that establishes and publishes standards for measurement, nomenclature and product performance in the UK. BSI is also the UK member body of the International Standards Organisation (ISO).

**BSN** Back-end Storage Network. Normally having a high bandwidth, this network is able to connect a number of computers to high capacity memory storage devices.

**BT** (1) British Telecom. An autonomous entity responsible for telecommunications in the UK and dealing with services such as telephone, telex, telegraph, Datel and Prestel. In the last quarter of 1984 BT was privatized.
(2) Busy Tone. A signal conveyed to a telephone calling party to indicate that the called telephone party or intermediate apparatus is reserved for use or is rendered unavailable for traffic.

**BTAM** Basic Telecommunications Access Method. BTAM is used to control the transfer of data, in a computer system, between the main storage and local or remote terminals.

**BTP** Batch Transfer Program. Able to accommodate more than one terminal, a BTP controls data transfers from local or remote computer terminals, and allows for modification or deletion of records.

**BTX** BildschirmTeXt. The West German viewdata system.

**BV** Breakdown Voltage. The reverse bias voltage at which a p-n (p-n) junction begins to conduct a large current. The junction will not be damaged unless excessive currents are allowed to flow. Compare with BDV which applies generally.

**BW** BandWidth. (a) A range of frequencies that can be occupied by a transmitted modulated signal. (b) The range of frequencies within which the performance of a device or system falls within given limits, with respect to a particular specified characteristic.

**BYTE** A colloquial expression to describe an 8-binary digit (BIT) word, in contrast to a 'nibble', which is generally considered to be a 4-binary digit word.

# C

**c** Centi-. A decimal prefix commonly used in association with a base unit in the SI system. Centi- denotes a multiplication factor of 0.01 or $10^{-2}$.

**C** (1) A high-level and structured computer programming language, which has been designed to optimize run time, size and efficiency.

(2) The hexadecimal (HEX) symbol for binary 1100, decimal 12. $C_{16} = 12_{10} = 14_8 = 1100_2$

(3) One of the letter designations given to the most commonly used sizes of 1.5-volt (V,1) batteries. Originally USA nomenclature, this type of letter description is gaining increasing popularity over the International Electrotechnical Commission (IEC) letter/number system as it is able to simplify the selection of batteries. *See* battery-size cross-reference guide, Appendix.

(4) Candle. Now obsolete, the international candle was a unit of measurement of light intensity and has been superseded by the candela. 0.982 international candles = 1 candela.

(5) Capacitance. In a system of plates (conductors) and dielectric (insulators), capacitance is the ratio quantity of electric charge to the potential developed. It is measured in farads (F).

(6) Capacitor. A component consisting of two electrodes or plates separated by a dielectric (an insulator). It has appreciable capacitance (5) above). The symbol C is used to represent a capacitor on parts lists and to annotate schematic diagrams.

(7) Carbon. A conductive non-metal which is also known as 'graphite' and has been used extensively in the electronics and electrical industries.

(8) Celsius. The suffix °C is sometimes taken to be degrees centigrade but now represents degrees Celsius. The two temperature scales are substantially the same, but with subtle differences. The centigrade scale is defined by 0° at the freezing point of water and 100° at its boiling point. The Celsius scale is defined to be the same as the Kelvin (K,2) thermodynamic scale with the zero shifted to the ice point, which is 273.15 K.

(9) Coulomb. The SI unit of electrical charge equal to the quantity of electricity transferred by a current of 1 ampere (A) flowing for 1 second (s).

**CAAS**  Computer-Assisted Acquisition System. A computer system used to assist libraries in their acquisition of material.

**CAD**  (1) Computer-Aided Design. Refers to the capability of a computer to be used for automated industrial, statistical or architectural design by the use of visual devices.

(2) Computer-Aided Dispatch. A system in which hand written incident cards and conveyor belts in radio dispatching operations are replaced by alphanumeric keyboards (KB) and cathode ray tube (CRT) terminals.

**CADC**  Computer-Aided Design Centre. Located near Cambridge in England, this installation's major function is computer aided design (CAD).

**CAD/CAM**  Computer-Aided Design/Computer-Aided Manufacture. A method of production employing designs provided by a computer aided design (CAD) system held in a memory device, to produce finished products by computer or numerical controlled (N/C) machinery.

**CAD/CAT**  Computer-Aided Design/Computer-Aided Test. A method of production employing designs provided by a computer aided design (CAD) system held in a memory device, to provide final testing of a finished product under the control of a computer. The tests are most often undertaken by automatic test equipment (ATE).

**CADMAT**  Computer-Aided Design Manufacture And Test. A three-year scheme introduced in 1982, under the auspices of the UK Department of Trade and Industry, to promote the greater use of appropriate computer-aided techniques within the electronics industry. CADMAT techniques have been developed to satisfy the requirements of computer aided design/computer aided manufacture (CAD/CAM) when applied to the production of computers and electronic equipment, as well as additional specialist requirements for the design, verification layout and inter-connection of circuits, and for manufacture, procurement and testing.

**CAFS**  Content-Addressable File Store. Attachable to a mainframe computer as a peripheral, this stand alone unit permits a very high rate of searching of a structured set of records or database using various sets of fields or keys to identify a particular record.

**CAI**  (1) Computer-Aided Instruction. Enabling a student to converse with a computer, CAI uses a programmed study course that selects topics according to the user's response. Thus students are allowed to progress at their own rate of learning. (2) Computer-Assisted Instruction. This application of computer instruction provides for a dialogue between a student and a computer program, that informs the user of mistakes as they are made.

**CAIRS**  Computer-Assisted Information Retrieval System. A computer system that basically provides information to users in response to their requests. The computer has files containing information that can be searched for particular items, in response to the definition of particular user needs.

**CAL**  Computer-Aided Learning. The use of a computer in a system that is employed to aid students in a course of programmed learning.

**CAM**  (1) Computer-Aided Manufacture. A manufacturing production process resulting in finished products, controlled by a computer or numerical controlled (N/C) machinery.

(2) Content-Addressable Memory. A type of memory that can be addressed by its contents or meaning rather than by memory position or location.

**CAMAC** Computer-Automated Measurement And Control. A standard that defines the physical sizes of modules and racks, as well as defining power supplies, module address lines, data lines

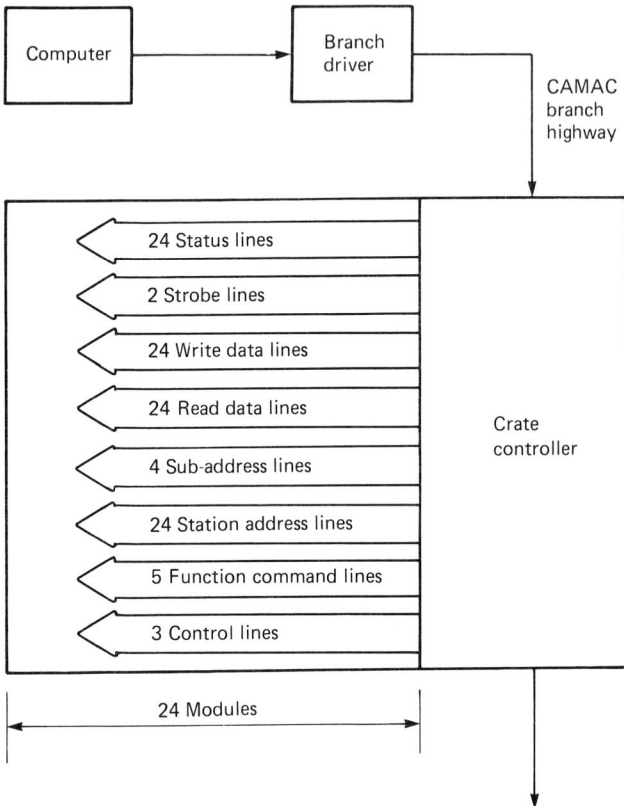

*Figure C.1* Configuration of the computer-automated measurement and control (CAMAC) branch highway

and control lines. This in-depth specification (SPEC) is used for the general interfacing of instruments and was originally developed for the nuclear industry to transfer large volumes of data quickly with no errors. *See Figure C.1.*

**CAN** CANcel. (a) The cancel character instruction is used to indicate data that is in error and should not be accepted. (b) A control code for the deletion of information used in American standard code for information interchange (ASCII).

**CANCL** CANCeL. This status word is used to indicate that a remote system has deleted information previously transmitted.

**CAR** (1) Channel Address Register. A storage location that provides information to access a path to a particular register, location or unit of memory where information is stored.

(2) Computer-Aided Retrieval. The use of computers in document-retrieval systems (DRS) to replace the simple manual look-up reference systems.

**CARR** CARRier. An unmodulated radio frequency (RF) source signal of a specific wavelength, usually modulated by frequency modulation (FM) or amplitude modulation (AM) variations, representing the information to be conveyed.

**CAS** Column Address Strobe. Used in dynamic memory control to find the position of a word in a memory.

**CATV** CAble TeleVision. Previously known as 'community antenna television', CATV refers to the use of coaxial cable (COAX,2), or various other cable loops to deliver television (TV) signals to subscribers. Sometimes also known as 'piped aerial'.

**CAV** CAVity resonance. A cavity formed by a conducting surface provides this effect, its size and shape being chosen to give resonance to a particular frequency of electromagnetic wave.

**CAW** Channel Address Word. A word that provides an access path to a particular register, location or unit of memory where information is stored.

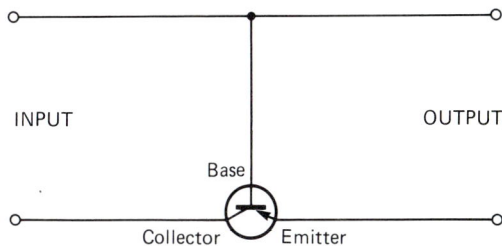

*Figure C.2* Basic form of a common-base (CB,2) connection bipolar-transistor amplifying circuit

**CB** (1) Citizen Band. A frequency band allocated to privately-owned stationary or mobile transmitter/receivers, that are intended for operation by unskilled operators.

(2) Common Base. A particular type of transistor amplifier

(AMP) that has its base element common to the input and the output (O, 2) circuits. *See Figure C.2.*

(3) Central Battery. Employed in a central battery telephone system, in which energy for signalling and speaking is drawn from a power installation at the exchange. Also called a 'common battery', used in a common battery system.

**CBI** Common Batch Identification. A method by which a number of groups of computer records or documents can be recognized.

**CBL** Computer-Based Learning. A structured course of teaching and learning that relies on the use of a computer.

**CBMS** Computer-Based Message System. A term used to describe the communication via electronic mail of keyboarded material which also includes the idea of electronic mailboxes.

**CBX** Computer-controlled Branch eXchange. Controlled by a computer, this type of telephone exchange is able to offer various additional facilities not normally available on an ordinary non-intelligent telephone exchange.

**CC** (1) Common Collector. A particular type of transistor amplifier (AMP) that has its collector element common to the input and output (O, 2) circuit. *See Figure C.3.*

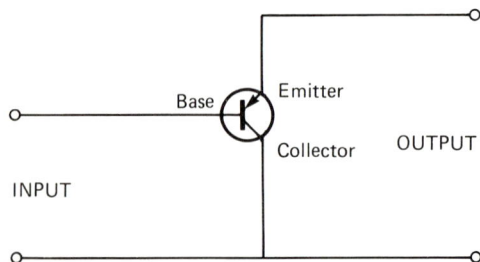

*Figure C.3* Basic form of a common-collector (CC,1) connection bipolar-transistor amplifying circuit

(2) CalCulator. (a) A data processor suitable for performing arithmetical operations with little human intervention. (b) A device for carrying out arithmetic, digital and logic operations.

(3) Charge Conveyor. A device or conducting path capable of transferring a flow of electricity.

(4) Chip Carrier. A high-density package for very large scale integration (VLSI) integrated circuits (IC). It may be a leadless chip carrier (LCC), a leadless ceramic chip carrier (LCCC), a pin grid array (PGA) or a leadless land grid array (LLGA). There are many different sizes available.

**CCB** Channel Control Block. A set of locations in memory containing information that is accessed by various programs and provides control of the input or output (I/O) of data via a channel.

**CCCCD ($C^4D$)** Conductivity-Connected Charge-Coupled Device. A hybrid between a charge-coupled device (CCD) and a bucket-brigade device (BBD). It is a CCD that employs doped regions between its potential wells.

**CCD** Charge-Coupled Device. A unit that is basically a charge-transfer device (CTD), consisting of metal-oxide semiconductor (MOS) capacitors (C, 6), as shown in *Figure C.4*.

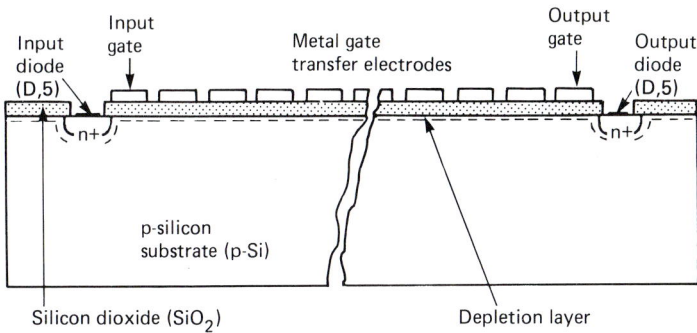

Input diode (D,5)
Input gate
Metal gate transfer electrodes
Output gate
Output diode (D,5)
n+
n+
p-silicon substrate (p-Si)
Silicon dioxide ($SiO_2$)
Depletion layer

*Figure C.4*  A cross-sectional view of a charge-coupled device (CCD)

The capacitors are coupled so that charges can be moved through the semiconductor substrate in a controlled manner. These devices can perform a variety of functions including serial memories, signal processing and dynamic filtering.

**CCETT** Centre Commun d'Études de Television et Telecommunications. A television and telecommunications research establishment based in France.

**CCIR** Comité Consultatif International des Radiocommunications. (Consultative Committee on International Radio.) One of the three main organizations within the International Telecommunications Union (ITU), particularly involved in the examination and standards recommendation of long-range radio communications.

**CCITT** Comité Consultatif International Télégraphique et Téléphonique. (International Telegraph and Telephone Consultative Committee.) One of the three main organizations within the International Telecommunications Union (ITU), concerned with devising and proposing recommendations for international telecommunications.

**CCP** Console Command Processor. An area in the memory map (MAP, 2) of a microcomputer (MP) using a control program for microcomputers (CP/M) disk operating system (DOS) that allows the user to converse with the operating system by using the command syntax. Thus the user can accomplish various tasks which include listing a directory on a disk or executing a program file on a disk simply by using the terminal that is tagged on the system.

**CCS** Common-Channel Signalling. The use of a specific channel for the transmission of information relating to the control of calls in other channels. The channel transfer of network management information.

**CCSA** Common-Control Switching Arrangement. An American term for a switching facility to corporate tie-line networks. The switching of leased lines in an organization's network is then accomplished by switching equipment. Thus, stations in the network can dial each other without charge, and outside calls can be made by dialling, normally 8 in the USA and 9 in the UK.

**CCT** CirCuiT. A conducting path comprised of electrical or electronic devices which together perform a desired function.

**CCTA** Central Computer and Telecommunications Agency. Part of a UK Civil Service department, having responsibility for policy on government administrative telecommunications requirements. It manages the Government Telecommunications Network (GTN) and advises on public purchasing policies in computer and telecommunications fields.

**CCW** (1) Counter ClockWise. American terminology for 'anti-clockwise'.

(2) Channel Command Word. A set of binary digits (BITs) treated as a single unit which is transmitted to a channel to indicate the functions that it should perform.

**cd** CanDela. The SI unit of measurement of luminous intensity. One candela emits $4\pi$ of light flux.

**Cd** CaDmium. A metal having the atomic number 48.

**CD** (1) Current Density. In a current-carrying medium which can be either a radiation beam or a conductor, CD is the ratio of current to cross-sectional area of the medium.

(2) Compact Disk. An audio disk system in which the disk is only one seventh of the total area of a conventional record or disk, although a single side can play for an hour. Information is carried or encoded as spiral groove of pits and flats in digital (D,2) form. The CD carries a total of over five billion digital sound signal binary digits (BITs) not including additional bits that are used for speed control and error correction tasks.

(3) C or D. (a) Special paper of uniform size and shape which has holes or various marks or arrays that can be sensed by brushes or photoelectric pick ups. (b) A pluggable unit for printed circuit wiring and components.

**CDC** (1) Code Directing Character. One or more routing indicators, used in telecommunications at the start of a message to determine its destination.

(2) Call Directing Code. An identifying call, usually two letters in length, that is transmitted to an outlying receiver to turn on its printer automatically.

(3) Control Data Corporation. A US-based manufacturer of computers, in particular large processing systems.

**CDH** Command and Data Handling. Refers to the decoding or storing of an electronic pulse signal or set of signals (command) and basic elements of information (data).

**CDM** Colour-Division Multiplexing. In the visible region of the electromagnetic frequency spectrum, CDM is a similar process to frequency-division multiplexing (FDM). Each colour corresponds to a different frequency and a different wavelength.

**CDP** (1) Checkout Data Processor. A device that has the capability of performing the reduction, summarizing, processing, or input of information and has a set of routines developed to provide a programmer with an evaluation of a program under operating conditions.

(2) Communications Data Processor. A device that has the capability of performing the reduction, summarizing, or input and output, of transmitted data originating from another point.

**CDPM** Coloured Digital Panel Meter. A meter providing the usual features of conventional panel meters and in addition changes the colour of the display at desired input levels set by trimmer potentiometers (POT). Typically at the lowest input the display is green, it then changes to yellow, and then to red as the input to the meter is increased.

**CDT** Control Data Terminal. A point at which items of data that are used to identify, select, execute or modify, may be entered or can leave a communications network.

**CDU** Central Display Unit. A device (usually a VDU) providing a visual representation at a single location of data obtained from a number of geographical locations.

**CE** (1) Common-Emitter. A transistor amplifier (AMP) in which the emitter element is common to the input and the output (O, 2) circuit. *See Figure C.5.*

(2) Customer Engineer. Employed by the original equipment manufacturer (OEM), this engineer is responsible for the maintenance of the system when it is installed at the customer's premises.

31

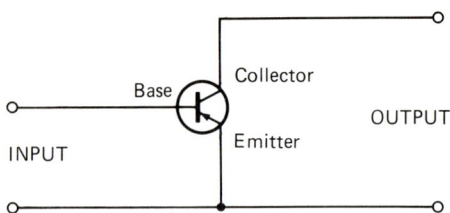

*Figure C.5* Basic form of a common-emitter
(CE,1) connection bipolar-transistor amplifying
circuit

**CECC** CENELEC Electronic Components Committee. Evolved from the BS 9000 system (BS 9000), the CECC was formed by eleven European nations in 1970 to establish a similar system of quality assurance over a wide range of components under the authority of the European Committee for Electrotechnical Standardisation. The system has the objective of harmonizing specifications and quality assessment procedures for electronic components, by the granting of an internationally recognized Mark and/or Certificate of Conformity. The member countries are Belgium, Denmark, France, West Germany, Ireland, Italy, Netherlands, Norway, Sweden, Switzerland and the United Kingdom. (*See* CENELEC).

**CEE** International Commission for the Approval of Electrical Equipment. Comprising the national electrotechnical committees of 22 European countries, the CEE founded in 1946 also includes as observer members Australia, Canada, Hong Kong, Iceland, India, Spain, Japan, South Africa and the USA.

**CEEFAX** BBC's teletext. A play on words from 'see' and 'facts'. A system which transmits data in the vertical blanking interval of a television (TV) picture. The data can be accessed only by users who have a special decoder unit.

**CEN** European Committee for Standardisation. CEN comprises the national standards bodies of 14 European Economic Community (EEC) and European Free Trade Association (EFTA) countries along with Greece and Spain. It publishes European Standards which have to be accepted by a large majority of its members and are published without any variation of text.

**CENCER** European Committee for Standardisation Certification. The certification body of the European Committee for Standardisation (CEN).

**CENELEC** European Committee for Electrotechnical Standardisation. The electrotechnical counterpart of the European Committee for Standardisation (CEN). It was formed from a union of CENEL and CENELCOM and comprises the national electrotechnical committees of 15 West European countries as well as over 30 technical committees.

**CEOC** Colloque Européene des Organisations de Contrôle. Comprising leading independent technical inspection organisations in most European countries, CEOC was formed in 1961. Its aim is to harmonize the practical standards for the inspection of engineering plant and machinery, by the exchange of information.

**CEPT** Conference Européene des Administration des Postes et des Télécommunications. (Conference of European Postal and Telecommunications Administrations.) A European conference body that makes recommendations for telecommunications practice and standards within Europe.

**CERDIP** CERamic Dual-In-line Package. A dual-in-line package (DIP) with high reliability characteristics and a very durable and rugged form of housing.

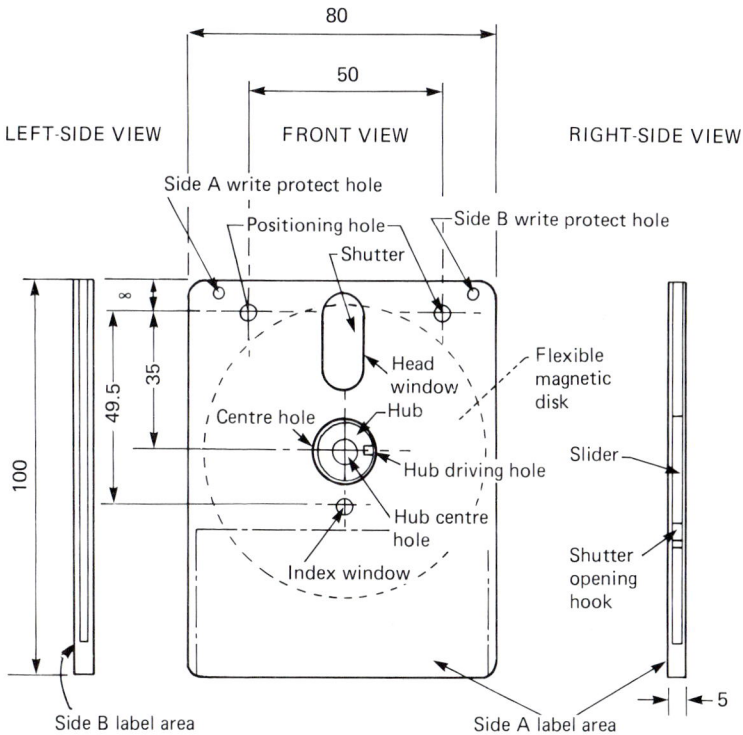

*Figure C.6* Component parts and dimensions (in millimetres) of a compact floppy disk (CFD)

33

**CERN**   Centre Européene pour la Recherche Nucleaire. Based on the Swiss border of France, this organization is concerned mainly with nuclear physics research.

**CF**   (1) Control Footing. In computing, a minor total that is a summary at the end of a control group or of various control groups.

(2) Central File. May be thought of as a single electronic reservoir or data-deposit, containing a central information file, index and all major application files.

(3) Count Forward. The act of a device adding so that it counts the increases of a common cumulative total of appearances or occurrences of a particular event.

**CFD**   Compact Floppy Disk. A smaller form of floppy disk (FD,1), being 3, 3¼ and 3½ inches rather than the conventional mini disk sizes of 5 and 5¼ inches.

CFDs have a rigid hard plastic shell measuring 80 × 100 × 5 mm (*see Figure C.6*) with a head window shutter on the case that automatically opens and closes when the case is inserted or pulled out of a compact floppy disk drive (CFDD) system.

**CFDD**   Compact Floppy Disk Drive. CFDDs have been designed to match and drive each size of compact floppy disk (CFD) and are direct, plug-compatible replacements for conventional mini floppy disk (FD,1) drives. The rotation speed, data transfer rate, recording capacity per track and other specifications are exactly similar to those of the mini floppy disks, so existing floppy disk controllers (FDC) can be employed.

**CFR**   Catastrophic Failure Rate. The number of times any type of failure occurs that renders the useful performance of a computer or equipment to zero.

**CGB**   Convert Gray to Binary. The process of changing numerical information from Gray (a type of code wherein, passing from one decimal digit to the next sequential digit, only one binary digit changes its value) to Binary (a coding system in which the encoding of any data is done through the use of binary digits (BITs), i.e. 0 or 1).

**CGS**   Centimetre, Gram, Second system. A system of units in which the fundamental units of length, mass and time are the centimetre, gram and second respectively (now generally obsolete).

**CH**   Control Heading. The title or short definition of a control group of records which appears in front of that group.

**CHG**   CHanGe. An alteration to master files or computer programs.

**CHILL**   CCITT HIgh-Level Language. A high-level computer language used particularly for real-time applications such as the control of electronic telephone exchanges, put forward by the Comité Consultatif International Télégraphique et Téléphonique (CCITT).

**CHIP** (1) A colloquial expression for capacitors (C,2), resistors, microprocessors (MP) or other devices manufactured on small pieces (chips) of silicon, ceramic or other materials (i.e. 'chip resistor').

(2) Chip Hermetically In Plastics. Microprocessors (MP) or other devices cut from a larger wafer, usually silicon, hermetically sealed within plastics or resin.

**CHKPT** CHecKPoinT. A point of time during a machine run where processing is halted momentarily to allow the variables of the run to be recorded on magnetic tape (MAGTAPE).

**CHNL** CHaNneL (a) A particular path or specific frequency band in communications used for the reception or transmission of signals. (b) In a computer (CMPT) or data-processing system (DPS, 1) a route along which information travels. (c) In data transmission, a means of performing two-way-alternative communication in one preassigned direction. (d) An individual path in a transmission medium, that is separated from other paths, e.g. a multipair cable. (e) A connection between two nodes in a network. (f) In communication theory, the part of a communication system that connects a data source to data acceptacne equipment. (g) A complete facility for the transmission, emission or reception of signals, e.g. a television (TV) channel or a radio channel. (h) In a field-effect transistor (FET), the region connecting a current source to the point where charge carriers leave.

**CHRG** CHaRGe. An excess or deficiency of electrons, which provides a body with negative or positive electrification, i.e. a quantity of unbalanced electricity in a body.

**CI** Call Indicator. A flagging device in a computer system that shows when branching or transfer of control to a specified closed subroutine has taken place.

**C/I** Carrier-to-Interface ratio. A specified ratio between a carrier wave on either side of a boundary between two systems or devices.

**CICS** Customer Information Control System. A system that provides an interface between operating systems (OS), access methods and applications programs to allow remote or local display terminal interaction with the data base of the central processor.

**CIF** (1) Central Index File. (a) A centrally located table containing a set of key characters, treated as a whole, that identify the actual disk records in another permanent disk file.

(b) A central file of terms in an automatic retrieval system that is searched by a computer.

(2) Central Integration Facility. A centrally located device whose output (O,2) function is proportional to the integral of the input function with respect to a specified variable.

**CIM** Computer Input Microfilm. A system whereby images on microfilm are scanned and converted into a digital form and then stored in the memory of a computer.

**CIO** (1) Central Input Output multiplexor. A centrally-located device that takes low-speed inputs or outputs (O,2) from terminals and combines them into one high-speed data stream, for transmission between an external source and a computer.

(2) Carrier Insertion Oscillator. In single sideband (SSB) transmission, a CIO is used to replace the suppressed sideband at the receiver to permit demodulation.

**CIOCS** Communication Input/Output Control System. A group of routines which automatically control the performance of input/output (I/O) operations an direct other functions, in a computer system that handles on-line, real-time applications (i.e. a communications system).

**CIP** Cataloguing In Publication. Data that is held in a National Library's machine-readable cataloguing (MARC) files, providing details on material that is in the process of publication.

**CIR** CIRcuit. A combination of electrical or electronic devices forming a conducting path, that together carry out a desired function.

**CISPER** Comité International Special de PEtulsation Radioélectriques. An international body formed to set specific limits, test equipment specifications (SPEC), test procedures and electromagnetic interference (EMI) controls throughout most of the world.

**CISPR** International Special Committee on Radio Interference for the Approval of Electrical Equipment. Formed under the authority of the International Electrotechnical Commission (IEC), the CISPR brings together IEC member committees and other international organizations in the electrical and broadcasting fields.

**CIT** Call-In-Time. A period of time in which the control of digital computer is transferred from a main routine to a sub-routine. The period has been inserted into a sequence to fulfil a subsidiary purpose.

**CIU** Computer Interface Unit. A unit that provides a common boundary between automatic data processing systems (ADPS). In communications and data systems it may involve code, format, speed or other changes as required.

**CK**  ChecK. (a) A means of verifying the accuracy of data transmitted, manipulated or stored by any unit or device in a computer system. (b) A process of testing the corrections of machine operations, the existence of certain prescribed conditions within a computer, or the correctness of the results produced by a program.

**CKDIG**  ChecK DIGit. (a) In a character or word, one or more redundant digits which depend on one another such that if one digit changes, a malfunction can be detected. (b) When carried along with a machine word, one or more digits used in relation to other digits in the word as a self-checking or error-detecting code.

**CKO**  ChecKing Operator. The what-to-do portion of a checking operation.

**CKT**  Circuit. A combination of electrical or electronic devices forming a conducting path, that together carry out a desired function.

**CLK**  CLocK. Used to time events within a piece of equipment or device, a clock provides a timing signal of a precise frequency. Normally crystal-controlled, it is the basic method of producing periodic signals for synchronizing electronic equipment, especially computers.

**CM**  Communications Multiplexor. A communications device that provides for the transmittion of two or more messages simultaneously over the same channel (CHNL, (a)). It achieves this by frequency-division multiplexing (FDM), or by time-division multiplexing (TDM).

**CMA**  Circular Mil Area. A unit of area used to indicate a wire size. One circular unit equals the cross-sectional area of a wire one mil (0·001 inches) in diameter.

**CMD**  CoManD. (a) An electronic pulse, signal or set of signals to start, stop or continue an operation. (b) The portion of an instruction word which specifies the operation to be performed.

**CML**  Common Mode Logic. A family of integrated logic circuits, with pairs of transistors coupled by their emitters forming a fundamental part of a circuit.

**CMOS** or **C/MOS**  Complementary Metal-Oxide Semiconductor. A pair of metal-oxide transistors of opposite type used together, as shown in *Figure C.7*, CMOS uses the complementary symmetry of n-channel and p-channel transistors to switch a voltage from one circuit to another with very low power dissipation or loss.

**CMPT**  CoMPuTer. An automatic device capable of accepting information, applying prescribed processes to the information, and supplying the results of these processes. The instructions

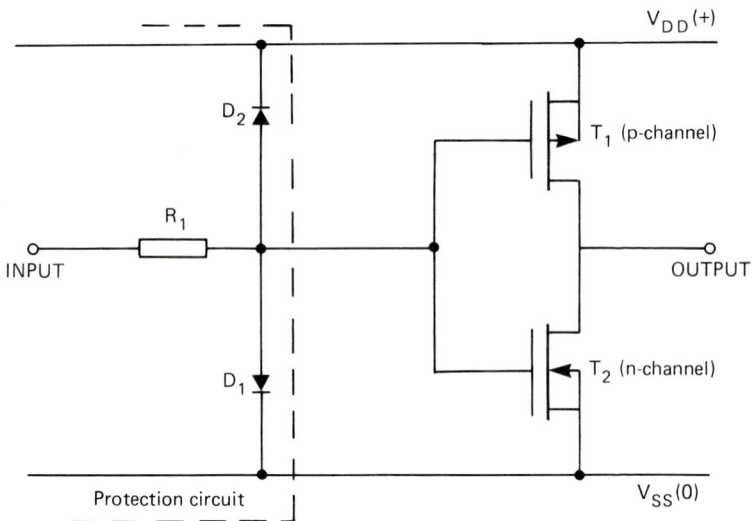

*Figure C.7* A complementary metal-oxide semiconductor (CMOS) inverter circuit

are stored in a memory. The most widely used and most versatile of these devices is the digital computer.

**CMPLX** CoMPLeX. Usually refers to programs or circuits which have increased intricacies of structure or design, which produce significantly faster devices.

**CMR** (1) Common-Mode Rejection. An ideal differential amplifier (AMP), produces an output of zero when the inputs are identical. In practice a small positive or negative signal may occur, known as common-mode rejection.

(2) Communications Moon Relay. A relay station on a moon landing module, used for communication between Earth control and astronauts mobile on the surface of the Moon.

**CMRR** Common-Mode Rejection Ratio. The measure of the ability of a differential amplifier (AMP) to produce a zero output (O,2) for like inputs. *See* CMR,1.

**C/N** Carrier-to-Noise. Expressed in decibels (dB), C/N is the ratio of the value of a carrier wave, to that of unwanted electrical disturbance (noise) after detection.

**CNC** Computer(ized) Numerical Control. A numerical control system that uses a dedicated stored computer program to perform some or all basic numerical control functions. Usually used to control machine tools such as lathes, etc.

**CNCT** CoNneCT. The action of using a device that enables

38

electrical or other impulses to be transferred between two pieces of equipment.

**CND** CoNDition. In the common business-orientated computer language (COBOL), this is an expression of a set of specified minimum or maximum control values that a specified datum can assume.

**CNR** Carrier-to-Noise power Ratio. The ratio of the value of a signal wave to that of unwanted electrical disturbance (noise) after detection.

**CNT** CouNTer. (a) Any electronic circuit, device, sub-assembly or instrument that counts electronic pulses. (b) A device that detects and counts individual particles and photons. (c) A register or storage location in a computer or an electromechanical unit forming part of another device, that is used to accumulate totals.

**CNTRL** CeNTRaL. Refers to a position in a network or system which is connected with, or is in communication with, all other points in the network or system.

**COAM** Customer Owned And Maintained. Pieces or sections of equipment that are exclusively maintained by engineers employed by the equipment owners.

**COAX** (1) COAX. A colloquial expression referring to a coaxial cable.
(2) COAXial cable. A cable consisting of concentric outer and inner conductors, separated from each other by insulation.

**COBOL** Common Business-Orientated Language. A computer language designed for commercial data processing (DP) such as accounting and financial work. It is a specific language by which business-data processing procedure may be precisely described in standard form.

**CODASYL** COnference for DAta SYstems Languages. Created by the US Department of Defense along with computer users and manufacturers, this group has specified a number of application-independent and manufacturer-independent software tools (e.g. COBOL).

**CODEC** COder-DECoder. An analogue-to-digital (A/D,2) and digital-to-analogue (D/A,2) converter. It is used to convert analogue signals such as speech, music or television to digital form for transmission over a digital medium and back again to the original analogue form. *See Figure C.8.*

**CODIC** COmputer-DIrected Communications. Transfer of information in various forms, routed via a computer from one point, person or device to another.

**COED** Computer-Operated Electronics Display. A form of display (D,4) at large sports stadiums or on advertisements

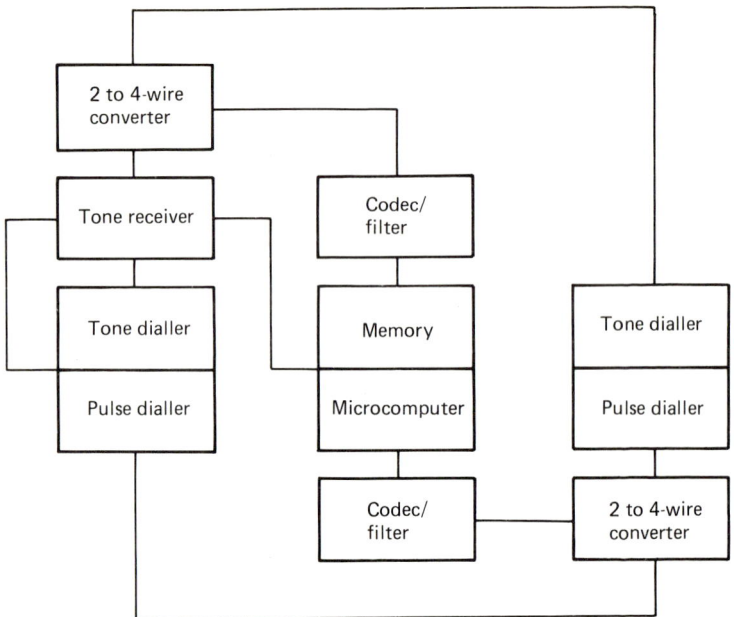

*Figure C.8* A typical design of a voice channel telecommunications coder-decoder (CODEC)

consisting of lights used to construct words or shapes, controlled by computer.

**COGO** Coordinate GeOmetry. A computer language that is used for solving coordinate geometry problems in civil engineering.

**COHO** COHerent Oscillator. A device that produces radiation waves having pure well-defined oscillations, such as in a laser.

**COL** Computer-Orientated Language. A programming language requiring little translation. This type of program usually runs efficiently on a related computer, but requires extensive translation for use on another variety of computer.

**COM** Computer Output Microfilm. A microfilm printer that will substitute directly for a line printer or tape output by taking output (O,2) directly from a computer. It may also be referred to as Computer Output Microfilmer, Computer Output Microform, Computer Onto Microfilm, Computer Output on Microfilm, Computer Output Microfiche or Computer Onto Microfiche.

**COML** COMmercial Language. Generally procedural, this type of computer language is designed for data processing (DP,1), accounting and financial work (*see* COBOL).

40

**COMM** COMMunication. The process of transferring information in various forms from one point, person or device to another.

**COMM-STOR** COMMunications STORage unit. A single- or dual-drive diskette system used for storing messages received on line or other locally-prepared information ready for transmission.

**COMPAC** Computer Output Microfilm PACkage. A package that substitutes directly for a line printer or tape output (O,2), being a microfilm printer that will accept output information directly from a computer.

**COMPEDEX** COMPuter Engineering inDEX. A computer file containing information covering all branches of engineering.

**COMPUNICATIONS** COMPUters and commuNICATIONS. This contraction is a colloquial term for the joint use of computers and communication systems.

**COMSAT** (1) COMmunications SATellite. An artificial unmanned satellite in Earth orbit that provides communication links between locations on Earth which can be widely separated. It accepts and relays suitably modulated signals between earth stations.
(2) COMmunications SATellite Corporation. A privately owned American corporation established by statute as the exclusive international satellite carrier and representing the USA in the International Telecommunications Satellite Consortium (INTELSTAT).

**COMTEC** COmputer Micrographic and TEChnology. Based in California, this is an association of manufacturers and users of computer output microfilm (COM) equipment.

**CONVT** CONVerT. (a) To transfer information from one number base to another. (b) To alter numeerical information from one number base to another.

**COPANT** Pan-American Standards Commission. Comprising the national Standards bodies of the USA and eleven Latin American countries, COPANT is a co-ordinating organization concerned with the regional implementation of International Organisation for Standardisation (ISO) and International Electrotechnical Commission (IEC) standards and recommendations.

**CoPs** COmponent Placement System. A technique that supplies chip capacitors (C,6) in a reeled format so that users may employ automatic placement. The capacitors are packaged in a continuous cardboard carrier strip, sandwiched between two layers of transparent tape with Electronic Industries Association (EIA) standard sprocket-drive holes.

41

**COSMOS** A synonym for CMOS or C/MOS.

**CP** (1) Clock Pulse. Used to control the timing of count characters, read circuits or other related functions, this pulse is located near recorded data.

(2) Critical Path. The overall time to complete a project cannot be less than that needed along the critical path. Any delay along this path can delay the whole project, whereas small interruptions along non-critical paths do not.

**CPC** Computer Power Centre. A power device that requires only one connection to the mains power, providing regulated power that can be distributed to various machines in a computer room. It also provides a computer with regulated power and protection from power disturbances.

**CPE** Central Processing Element. An element that can provide a complete 2, 4 or 8 binary digit (BIT) slice across the data processing (DP,1) section of a computer. It enables functions to be excuted in one, rather than several, microcycles.

**CPI** Characters Per Inch. A measure of the number of individual characters (including necessary spaces) that a printer, teletype-writer (TTY) or similar device can print along a line.

**CPL** Combined Programming Language. A very high-level language developed jointly in the 1960s by the University of Cambridge and the Institute of Computer Science.

**CPM** (1) Cards Per Minute. A measure of the rate at which a card reader is able to sense and read information punched on cards.

(2) Critical Path Method. Used for scheduling and controlling large projects, this technique of management is particularly useful for projects that involve a number of phases which depend upon one another. *See* CP,2, critical path.

**CP/M** Control Program for Microcomputers. An operating system (OS) written to run on 8080/8085/Z80 based micro-computers. It is the most widely used disk operating system (DOS) and a large number and variety of languages and utility programs have been written to run under CP/M. There are several versions, the most recent being 2.0.

**CPS** (1) Characters Per Second. (a) The number of individual letters, numerals or other symbols which are printed in one second. Usually this abbreviation occurs when discussing the speed of characters-at-a-time printers. (b) A measure of the rate of transmission normally between a computer and terminal devices. As various numbers of binary digits (BITs) are required to represent a character, there is no simple correlation between bits per second (BPS) and CPS.

(2) Cycles Per Second. Now superseded by hertz (Hz), CPS is

the unit of frequency equal to the number of complete waves per second.

**CPT**  Colour Picture Tube. A cathode-ray tube (CRT) designed to produce coloured pictures for television (TV). The image depends on the excitation of different phosphors to produce three primary colours, red, green and blue. The basic construction of a CPT is shown in *Figure C.9*.

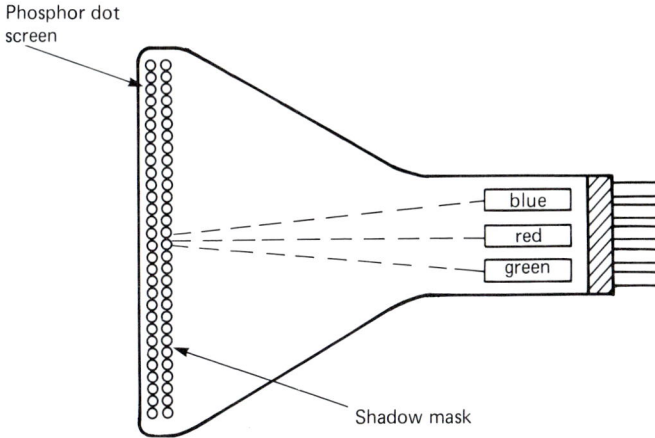

*Figure C.9*  The basic construction of a colour-picture tube (CPT)

**CPU**  Central Processor Unit. The portion of a computer that controls the operation of the whole system, where instructions are executed and computations performed. It contains the main storage, arithmetic unit and special register groups. Sometimes also known as Control Processor Unit.

**CPY**  CoPY. To reproduce data in a new location, replacing the information that was previously stored there without altering the source of the original data.

**CR**  Carriage Return. (a) The action of moving a carriage back to the beginning of a line on a printer or teletypewriter (TTY). (b) A key on many computer systems or teletypewriters (TTY), the operation of which returns the carriage or transmits a carriage return character, from which the computer transmits a line feed character which in turn generates a new line.

**CRC**  Cyclic Redundancy Check. A method of checking a transmitted bit stream for errors. The receiving station compares the transmitted value of information on blocks of data with its own computed value and if they are equal then no error has occurred.

**CRD** Constant Ringing Drop. A relay employed in burglar alarm systems which, when activated even momentarily, will remain in a closed or alarm state until reset. Often a key is required to reset the relay and turn off the alarm.

**CRJE** Conversational Remote Job Entry. Enabling users interactively to edit files, this method of running a computer does not allow them to interact with tasks submitted to lower priority processes.

**CRO** Cathode-Ray Oscilloscope. A device that permits the visual inspection of high-frequency waveforms which would be totally impossible to observe by any other means. Most CROs incorporate a cathode-ray tube (CRT), but, a few employ a flat liquid crystal display (LCD).

**CROM** Control Read Only Memory. Designed and microprogrammed to decode control logic, this special type of read-only memory (ROM) is an important component in several types of two-chip microprocessors (MP); one chip contains the CPU and the other a CROM.

**CRR** Constant Ringing Relay. A relay used in burglar alarm systems which, when activated even momentarily will remain in a closed or alarm state until reset. Often a key is required to reset the relay and turn off the alarm.

**CRT** Cathode-Ray Tube. A vacuum tube with a viewing screen as an integral part of its envelope, it converts electrical signals into a visible form. CRTs consist of an electron gun that produces an electron beam, a grid that alters the brightness by varying the beam intensity and a luminescent screen for the display (D,4). The electron beam is moved across the screen either by deflection plates or magnets. Focusing of the beam may also be achieved by electrostatic or electromagnetic means or by a combination of methods.

**CRTC** Cathode-Ray Tube Controller. A circuit or device employed to provide control for a cathode-ray tube (CRT).

**C/S** Cycles per Second. Now superseded by hertz (Hz), C/S is the unit of frequency equal to the number of complete waves or cyclic processes that occur every second.

**CSA** Canadian Standards Association. A body formed to set safety standards and qualify equipment.

**CSCR** Complementary Silicon-Controlled Rectifier. A specific type of silicon-controlled rectifier (SCR) that controls the current flows from anode to gate instead of from gate to cathode.

**CSECT** Control SECTion. A main sequence of instructions or data in a computer program, that can be transferred, deleted or replaced with a control section from other program segments.

**CSIPR**  Comité Special et International sur les Parasites Radioté-légraphiques. Set up by the International Electrotechnical Commission (IEC) this committee establishes standards for telecommunications equipment, in particular the control of radio frequency interference (RFI).

**CSIRO**  Commonwealth Scientific and Industrial Research Organisation. A research organisation based in Australia.

**CSL**  Computer-Sensitive Language. A programming language written so that it is totally or partially dependent on the type of machine used.

**CSMA**  Carrier-Sense Multiple Access. A formalized method by which communication is carried out in broadband networks, including local area networks (LAN), such as Ethernet.

**CSMA-CD**  Carrier-Sense Multiple Access-Collision Detection. A particular formalized method by which communication is carried out in broadband networks, including local area networks (LAN), such as Ethernet.

**CSMP**  Continuous Systems Modelling Program. A computer program that provides a digitally produced simulated analogue system.

**CSO**  Colour Separation Overlay. A superimposition technique used in colour television (TV). A colour area such as blue is selected and viewed by one camera, and the output of another camera is superimposed into the blue area of the first. The complete picture is then transmitted.

**CSR**  (Controlled Silicon Rectifier) Thyristor. 'CSR' represents 'thyristor' on parts lists and annotated schematics. *See* SCR.

**CSW**  Channel Status Word. An individual element of a table that indicates the status of input-output (I/O) channels.

**CTC**  Conditional Transfer of Control. A machine instruction that can cause a departure from the normal sequence of executing instruction in a computer.

**CTD**  Charge-Transfer Device. A semiconductor device that is capable of transferring discrete packets of charge from one position to the next. It can be used for the storage of charge over a short time, the time being, dependent on the material the device is made from.

**CTI**  Comparative Tracking Index. The maximum voltage at which a material can withstand 50 drops of a test solution, normally 0.1% by mass of ammonium chloride in distilled water, without any tracking occurring. The test voltages and therefore CTI should be divisible by 25.

**CTL**  Complementary Transistor Logic. The use together of a pair of transistors, in a logic circuit, of opposite type. For example a pair of n-p-n and p-n-p transistors, manufactured on the same integrated circuit (IC). Such elements are often constructed from complementary metal-oxide semiconductor (MOS) field-effect transistors (FET).

**CTNE**  Compania Telefonica National de España. The Spanish post, telegraph and telephone (PTT) administration authority.

**CTS**  Clear To Stand. This control line runs to a terminal from a modulator/demodulator (MODEM) indicating that a carrier is present and that data may be transmitted. An RS-232 standard (*see* RS-232C).

**CUG**  Closed User Group. A collection of computer system users who can treat one system as dedicated to their own particular purposes, having no access to facilities outside of those available to the group, and not permitting access to users who are not members of the CUG.

**CUTS**  Cassette User Tape System. A system, normally micro-computer-based, that employs any of a number of types of digital cassette recorders to record information from any user specified instrument, in formats that can be read by most types of terminal.

**CVD**  Chemical Vapour Deposition. A process for making optical fibres (OF) where silicon along with other glass-forming oxides and dopants are deposited at high temperature onto the inner wall of a fused silica tube. From this tube a long thin fibre is finally drawn.

**CVPO**  Chemical Vapour Phase Oxidization. A process for the production of graded index (GI) optical fibre (OF), which have a loss of less than 10 dB/km and a bandwidth greater than 300 MHz-km.

**CVT**  Constant Voltage Transformer. A transformer (XFMR) that can provide a substantially constant value of voltage (V,2) independently of the current (I) supplied to it.

**CW**  (1) Carrier Wave. The basis frequency of an unmodulated signal, being sinusoidal oscillation or electromagnetic wave. This wave is then varied or modulated in a desired manner. The term can also be used to describe a direct current (DC) which is intended for modulation.

(2) Continuous Wave. A transmitted radio or radar wave that maintains both constant amplitude and constant frequency.

**CWP**  Communicating Word Processor. A specialized desktop computer having copying and communications facilities. It can send information to and receive pages of text in seconds from another word processor by means of a network which allows rapid inter-office or site-to-site text communications.

**d**   Deci. A decimal prefix commonly used in association with a base unit in the SI system of units indicating a multiplication factor of $10^{-1}$ of that unit. For example, one decibel (dB) is equal to 0.1 bels.

Although this particular prefix is not one of the multiples that increase in steps of $10^3$ preferred by SI, it is used and limited to occasions where the recommended prefixes are inconvenient.

**D**   (1) One of the letter designations given to the most commonly used sizes of 1.5-volt (V,1) batteries. Originally a USA nomenclature this type of letter description is gaining increasing popularity over the International Electrotechnical Commission (IEC) letter/number system as it is able to simplify the selection of batteries. *See* battery-size cross-reference guide Appendix.

(2) The hexadecimal (HEX) symbol for binary 1101, decimal 13. $D_{16} = 13_{10} = 15_8 = 1101_2$.

(3) Density. (a) In printed-circuit-board (PCB) design or circuit design in general, the physical closeness of components or devices. (b) Applied to computers, the closeness of space distribution on storage media such as magnetic drums, magnetic tapes (MAGTAPE), or cathode-ray tubes (CRT).

(4) Digital. The use of particular whole numbers to represent quantities in calculations; on/off, open/closed and true/false are all dual-state conditions that are capable of expressing in digital form all information stored, transferred or processed.

(5) Diode. An electronic device with two electrodes, an anode and a cathode. Diodes are most commonly used as rectifiers. The symbol D is used to represent diode on parts lists and to annotate schematic diagrams.

(6) Display. A visible representation of information which can be produced through various devices such as cathode-ray tubes (CRTs), plasma and liquid crystal displays (PDs and LCDs) and light emitting diodes (LEDs).

**da**   DecA. A decimal prefix used in association with a base unit in the SI system of units indicating a multiplication factor of $10^1$ of that unit. For example, one decameter is equal to 10 metres.

Although this particular prefix is not one of the multiples that increase in steps of $10^3$ preferred by SI, it is used and limited to occasions where recommended prefixes are inconvenient.

**DA**   (1) Data Acquisition. The process undertaken by a data acquisition system (DAS), where data is collected at one computer or system location from various other locations. It is then converted into digital (D,4) information for operating, supervizory or accounting records.

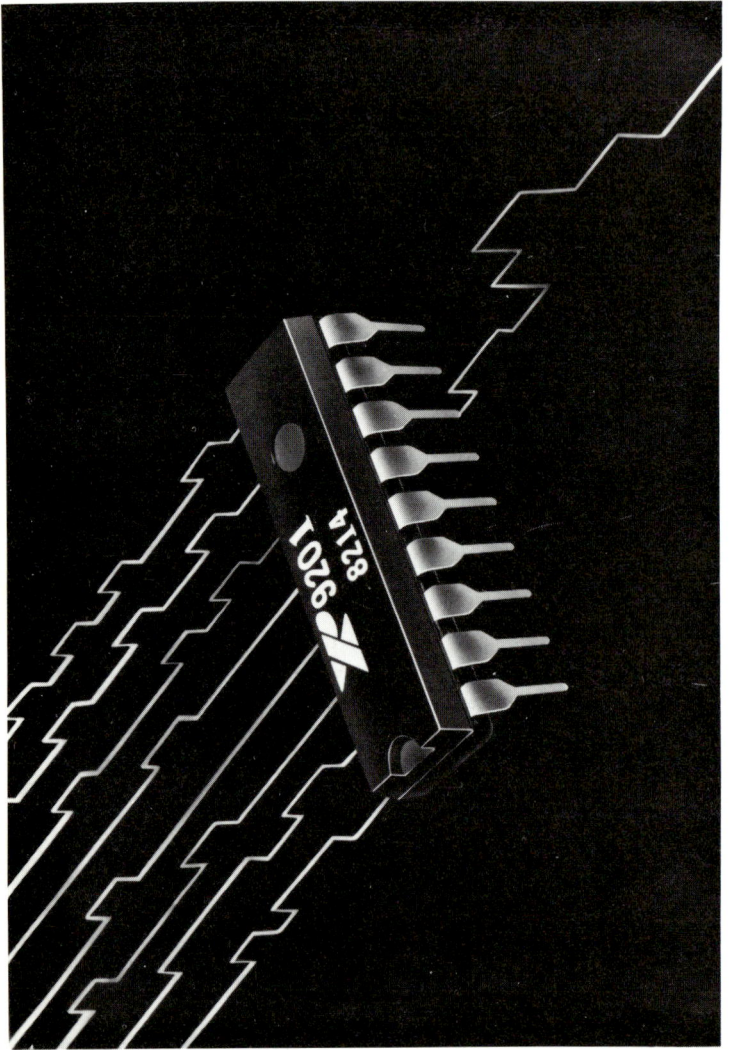

*Figure D.1* An 8-bit digital-to-analogue converter (D/A,2) (*courtesy* Thame Components Ltd)

(2) Direct Access. (a) Refers to the access or storage of data with reference to its own location and not dependent upon a previously accessed position. (b) A memory device that allows a memory location to be accessed independently of its position and information in various stores be accessed in the same amount of time.

(3) Differential Analyser. A device used for solving differential equations, usually an analogue computer, which employs interconnected integrators to solve the equations.

**D/A** (1) Digital-to-Analogue. The conversion of digital information and representation used in computers to the analogue signals necessary to drive speakers, meters and various other devices. The conversion is employed so that computers can talk, and operate with the outside world.

(2) Digital-to-Analogue converter. A unit which provides continuous analogue output (O,2) channels (CHNL(h)) from digital (D,2) input signals. *See Figure D.1.* Typically the outputs can range from 0 to +10 volts (V) and can be used as the control voltage for servo-mechanisms.

**DAA** Data-Access Arrangement. A small wall-mounted enclosure which contains an isolating transformer, that has been designed to prevent signals harmful to a direct-dial telephone network from being sent down the network lines. A DAA is required in the USA to protect the network from some customer-owned modulator demodulators (MODEMs). Also known as a 'direct-access arrangement'.

**DAC** (1) Digital-to-Analogue Conversion. The process of converting a digital signal to an analogue form. (Also D/AC).

(2) Digital-to-Analogue Converter. A unit that provides analogue output (O,2) channels (CHNL,1) from digital input signals. Typically the output can range from 0 to +10 volts (V) and can be used as the control voltage for servo-mechanisms.

(3) Data Acquisition and Control System. A wide variety of real-time (RT,1) applications, process control, and high-speed data acquisition (DA) can be handled by a DAC. Individually designed, each system is tailored to meet specific requirements.

**DACOM** DAtascope Computer Output Microfilmer. An early output (O,2) device used for computer output microfilm (COM) devices.

**DACS** Data Acquisition and Control System. *See DAC, 3.*

**DAM** Direct-Access Memory. A particular type of memory device that permits the access of a memory location independently of its position. The items stored can therefore be accessed in the same amount of time whatever their position.

**DAP** Distributed Array Processor. Attached as a peripheral to a mainframe computer, this device consists of a matrix of a large number of small processors that permit extensive calculations to be performed.

**DAR** (1) Damage Assessment Routine. A component of a computer operating system (OS) that endeavours to make a recovery from a software failure.

(2) Daily Activity Report. A report generated by computer concerning the daily uses and operation of programs, routines and subroutines.

**DARPA** Defence Advanced Research Projects Agency. An organisation in the USA which can provide support for the development of military and industrial projects such as rapid optical ocean surveillance testbed (ROOST) equipment.

**DAS** Data-Acquisition System. A system that collects or acquires data at one computer or system location from various other locations and then converts it into digital (D,4) information.

**DASD** Direct-Access Storage Device. A device that allows access time for any stored information on the particular medium to be independent of its location. Generally primary computer memory is direct, disks have pseudo-direct access and tapes have sequential access.

**DASM** Direct-Access Storage Media. Capable of storing information and computer programs, these media allow rapid access to individual elements. The access is independent of the location of the element required and of the location of the last element of data accessed.

**DATACOM** DATA COMmunications. (a) A general term used for devices or systems that transmit data from one point to another. (b) A communications network employed by the US Air Force covering the whole world. (c) A data communications service covering and linking more than 60 cities in the USA, set up by Western Union.

**DAV** (1) DAta Valid. Recognition given in a computer data entry system indicating that data entered has passed extensive arrays of data validation and error checking features.

(2) Data Above Voice. A system of digital (D,4) data transmission that carries information on a frequency above that used for voice transmission. *See* DUV.

**dB** DeciBel. A unit of measurement of the relative strength of a signal parameter, e.g. power or voltage.

**DBAM** Data-Base Access Method. A generic computing term used to describe various processes such as direct access (DA,2) and random access.

**DBI** Data-Base Index. An index used by the System Development Corporation (SDC), that provides information to all the files of computer information held by the organization. It is employed to help in the selection of a suitable database for a user requirements.

**dBM** DeciBel Milliwatt. Employed in communication work, decibel referenced to one milliwatt is a measure of absolute power. Zero dBM is equal to one milliwatt.

**DBMS** Data-Base Management System. The primary control software used in manipulating computer information files.

**DBOS** Disk-Based Operating System. Software contained in a computer which allows it to control the operation of all activities related to the use of magnetic disks in the system.

**dBrn** DeciBels above Reference Noise. Expressed as a ratio, dBrn is a comparison of the decibels above a certain reference noise to that reference noise. A reference noise is the magnitude of circuit noise that will produce a circuit noise meter reading equal to that produced by 10–12 watts (W) of electric power at 1000 hertz (Hz).

**DC** (1) Direct Current. Having a unidirectional flow this form of the movement of electric charge normally has a constant direction and a substantially constant magnitude. An example of direct current is that produced by a storage battery or cell.

(2) Data Collection. The process whereby data from one or more points is brought to a central position.

(3) Data Conversion. Essentially the process achieved by a hybrid or monolithic device that converts analogue data into digital (D,4) format or vice versa.

(4) Device Control. A character that does not form part of any message or information, this element of a character code is employed to effect control of the operation of equipment, by switching devices on or off. Usually used in telecommunication systems, but also sometimes used with other data-processing equipment.

(5) Direct Coupled. A particular design of circuit connected by resistance or inductance in such a manner as to permit direct current (DC, 1) to pass through the coupling.

(6) Data Channel. A data path which allows two-way communication between input/output (I/O) devices and a main memory in a computer.

(7) Double-Crucible. A process which uses two concentric crucibles to produce optical fibres (OF), as shown in *Figure D.2*: one for the glass forming the core and another for the glass forming the cladding. The complete fibre, including the cladding, is drawn from the bottom. Diffusion of dopant to each of the glasses produces a graded index (GI) fibre.

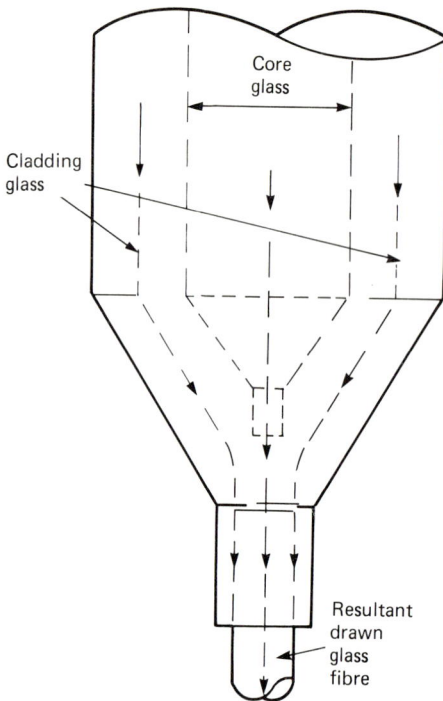

*Figure D.2* The double-crucible (DC,7) process for producing drawn optical fibres (OF)

**DCA** Defence Communications Agency. A US Department of Defense agency which now controls the packet-switched advanced research projects agency network (ARPANET).

**DCB** Data Control Block. A storage area holding control or status information employed by file management routines when storing and retrieving data.

**DCCU** Data Communications Control Unit. Messages that are received by a central terminal unit are scanned by the DCCU and transferred to the central processor unit (CPU).

**DC-DC** Direct Current to Direct Current. A DC-DC converter or regulator is a unit that is itself supplied from a direct current (DC,1) source providing a DC output (O,2). It can be employed in systems where a raw DC is available, giving a more stable DC output, plus isolation between input and output.

**DCE** (1) Data Communications Equipment. Usually used in reference to a modulator demodulator (MODEM), this term describes any piece of equipment that interfaces with a data communications network.

(2) Data-Circuit terminating Equipment. An active unit in a data station or terminal that connects a data transmission line to the terminal equipment. The interface unit controls all connections and provides signal or code conversions if required.

**DCMT** DeCreMenT. The progressive lessening or decay of a free oscillation due to the expenditure of energy. It is the ratio of two successive values of oscillation of the same sign.

**DCTL** Direct-Coupled Transistor Logic. A particular type of logic circuit that employs only direct-coupled (DC,5) transistors as active elements. *See Figure D.3.*

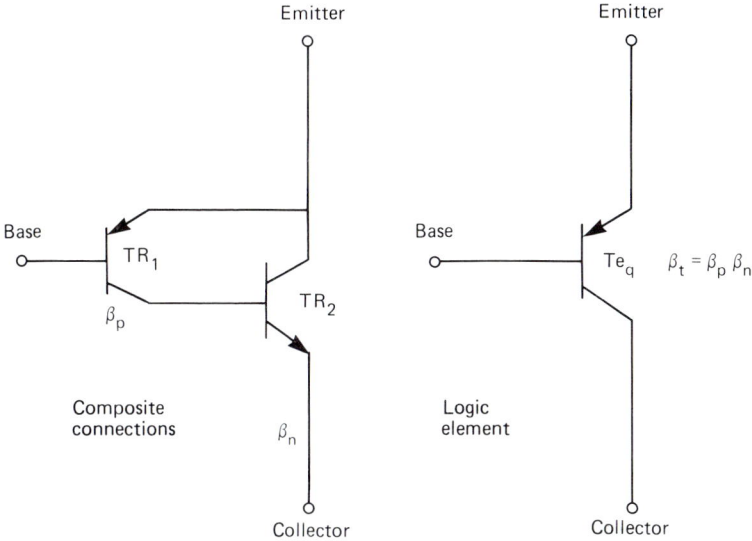

*Figure D.3* The composite connection of complementary transistors, forming a direct-coupled transistor logic (DCTL) element

**DCUTL** Direct-Coupled Unipolar Transistor Logic. A direct-coupled transistor logic (DCTL) circuit that employs field-effect transistors (FETs) as active elements (unipolar transistor being an obsolete term for field-effect transistor).

**DD** (1) Digital Display. (a) A display (D,6) of digital (D,4) orientated numerals. (b) Generally on a cathode-ray tube (CRT), a visual display of numeric, alphabetic or graphic material. (c) Information that is provided in a digital (D,4) form. (d) A display (D,6) of data in a decimal form.

(2) Data Definition. Commonly used in the job control language (JCL), where it is employed to connect external input/output (I/O) devices to an internal named file or data set (DS) associated with a particular step used in a computer program.

**DDA**   Digital Differential Analyser. (a) An incremental computer that employs a digital (D,4) integrator for computing. (b) A differential analyser that employs digital (D,4) representations for all analogue quantities used. (c) Having specific circuits designed to carry out mathematical integration, a DDA is part of a digital (D,4) computer.

**DDC**   Direct Digital Control. An automated action in a computer control system whereby the signal to a final control element is obtained by a digital (D,2) device. Such elements are used to perform process control tasks for production systems.

**DDCE**   Digital Data Conversion Equipment. Equipment employed in the conversion of digital data (DD,1(c)) into a form other than digital.

**DDCMP**   Digital Data Communication Message Protocol. Governing parallel, serial-synchronous, or serial-asynchronous data transmission and reception, this uniform discipline is used for the transmission of data between stations in a multipoint or point-to-point data-communication system.

**DDD**   Direct Distance Dialling. Information conveyed that automatically establishes, controls, supervises and releases a long-distance telephone call. It enables the telephone user to call outside his local area without operator assistance. In the UK and some other countries, this is called subscriber trunk dialling (STD).

**DDI**   Direct Dialling-In. Allows a telephone caller direct access to any particular public automatic business exchange (PABX) extension without involving the exchange operator. Also known as 'direct inward dialling' (DID).

**DDS**   (1) Dataphone Digital Service. An American communications service provided by the Bell System which eliminates the need for modulator demodulators (MODEMs) by transmitting data in a digital (D,4) form rather than an analogue one.
(2) Dewey Decimal System. A particular form of decimal classification system, developed by Melvil Dewey. It is one of the most widely used systems for cataloguing documents.

**DDT**   (1) Digital Data Transmitter. A package of computer software designed to assist in the isolation and correction of errors in a routine or program.
(2) Dynamic Debugging Technique. A computer technique that makes use of a debug program, especially designed for continuous debugging of user-written programs.

**DE**   Decision Element. In a computer, a DE circuit undertakes a logical function on binary digits (BITs) of input information and provides a result in its output (O,2).

**DEB**   Data Event Block. Memory control locations that have information accessed by programs, to provide control of asynchronous processing. The information is modified to indicate the progress of a particular event.

**DEC** DECimal. (a) A number-representation system with a number base of ten (i.e. the notation is represented by the characters 0 to 9). (b) An attribute or property which involves a condition, selection or choice from ten possibilities.

**DECIT** DECimal digIT. A particular character representing a whole number (not a fraction or mixed). In decimal (DEC) notation one of the characters 0 to 9.

**DECR** DECRement. The progressive lessening or decay of a free oscillation due to the expenditure of energy. It is the ratio of two successive peak values of oscillation of the same sign.

**DEF** Ministry of DEFence specifications. Until recently in the UK the main specification authority was the Ministry of Defence with their DEF specifications. However, these are gradually being superceded by the British Standard 9000 (BS 9000) series of standards for electronic components.

**DEL** (1) DELete. The action of removing or eliminating information, normally the removal of a record from a master file or the removal of a computer program from a memory after use.

(2) DELete character. A control character synonymous with back-arrow used to indicate that a previous character is to be ignored.

(3) DELay. (a) The time interval between the initiation of a signal and its reception; the time a pulse takes to traverse an electronic circuit or device. (b) In a computer the time taken to prepare or process data. (c) In data communications, the slowing down of information in a channel (CHNL(f)) for a specified period of time.

(4) Direct Exchange Line. Linked to only one subscriber, this single line has only one exchange number. Also known as 'exclusive exchange line' (EEL).

**DEM** DEModulator. Used in communications, a demodulator is

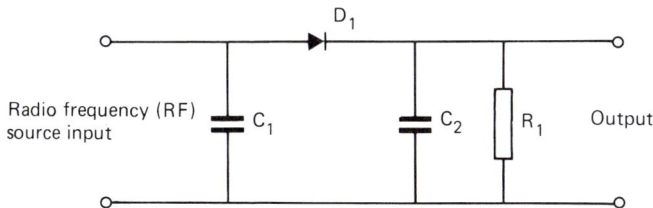

*Figure D.4*   A simple series-diode (D,5) demodulator (DEM) circuit

a circuit or piece of apparatus that extracts an original signal (or intelligence) from a modulated carrier wave with the minimum of distortion. *Figure D.4* shows a simple demodulator circuit.

**DEMOD** DEMODulator. *See* DEM, above.

**DEMUX** (1) DEMUltipleX. The division of one or more information streams into a larger number of streams.

(2) DEMUltipleXor. (DEMUltipleXer, US). A unit or piece of equipment that has a demultiplex ability of facility.

**DES** Data Encryption Standard. Used for the protection of sensitive data this security algorithm specifies implementation in hardware for maximum security. Data is passed through a complex circuit by means of a user-supplied key, outputting a cipher text that is a non-linear function of the original data which is then virtually impossible to regenerate without knowing the key.

**DESC** Defense Electronics Supply Center. An American term used for a centre that controls the procurement policies and monitors the quality of military electronics contracts.

**DETAB** DEcision TABle. Used for the analysis of various problems, a DETAB contains a list of the possible courses of action, alternatives or selections that can be made.

**DEU** Data Exchange Unit. Used in communications, a DEU accepts data from a number of input channels (CHNL (f)), and sorts it according to priority and destination, at the same time performing checking and error routines.

**DEUCE** Digital Electronic Universal Calculating Engine. Built in the UK this early computer superseded the Automatic Computing Engine (ACE).

**DEVCO** DEVelopment COmmittee. A council committee of the International Standards Organisation (ISO) aimed at the needs of developing countries.

**DF** (1) Double Frequency. A two-way radio device employing a separate and distinctly different frequency in each direction.

(2) Direction Finder. A system employed to determine the direction of origin of a transmitted signal. It uses a radio receiver and a directional antenna in combination.

(3) Diversity Factor. The ratio of the sum of the maximum demands from a group of consumers to their maximum collective demands.

**DFB** Distributed FeedBack. A type of injection laser, whose wavelength of radiative longitudinal-mode oscillations is determined by the corrugation period of the structure and not by its material composition.

**DFT** Diagnostic Function Test. A particular computer program employed to test the reliability of a system.

**DGT** (1) DiGiT. Any one of the symbols of integral value used in a system of numbering, each individual digit is used to designate a quantity smaller than the number base itself. In the decimal system the ten digits are 0, 1, 2, 3, 4, 5, 6, 7, 8, 9; in the binary system the digits are 0 and 1.

(2) Directorate General of Telecommunication. The postal, telegraph and telephone authority (PTT) of France.

**DH** Double-Heterostructure. A structure of injection laser, in which GaAsAl layers are fabricated on both sides of the gain region, which achieves reductions in the threshold current level. The device is capable of strong optical waveguiding and current confinement on both sides of its p-n junction (p-n).

**DHE** Data-Handling Equipment. A system of devices that can be either automatic or semi-automatic, used for the collection, transmission, reception and storage of information.

**DI** DeIonization. The process of removing dissolved salts or minerals from process water by ion exchange.

**DIANE** Direct Information Access Network for Europe. The information services offered over the European packet switching network (EURONET) system.

**DIB** Data Input Bus. Some microcomputers have a DIB, a mechanism in which data and address information is transferred between the switch register (SR) and the processor, also between processor and the memory, the memory and the input/output (I/O) interface and the processor and the input/output interface.

**DID** Direct Inward Dialling. A condition that allows a telephone caller direct access to any particular public automatic business exchange (PABX) extension without involving the exchange operator. Also known as direct dialling-in (DDI).

**DIDS** Domestic Information Display System. A computer graphics system that displays the whole of the USA in the form of colour-coded maps.

**DIGICOM** DIGItal COMmunications system. A collection of units and devices which transmit information from one point to another in a digital (D,4) form.

**DIL** Dual-In-Line. A concept that provides for the standardized location and connection of components and integrated circuits (ICs). It is based on two rows of connections spaced at 0.1 inch pitch. The term can be applied to the pins on a dual-in-line package (DIP) or to the holes in a socket or printed circuit board (PCB).

**DIMS** Distributed-Intelligence Microcomputer System. A method of multiprocessing which assigns to different processors fixed tasks allocated by software algorithms. The tasks include input/output (I/O) controller activities, data concentration, information processing, and remote communications.

**DIN** Deutsches Institut für Normung. (a) The West German specification system that can ostensibly be considered only as a guide to mating interface dimensions and tolerances. (b) Multipin audio and radio frequency (RF) connectors based on the West German standard are referred to as DIN connectors.

**DIP** Dual-In-line Package. A standard integrated circuit (IC) package. It has two rows of connection pins which extend down at 0·1 inch intervals. These packages are available in plastics for economy and ceramic for high humidities and temperature.

**DIV** DIVider. A circuit that reduces the number of pulses or cycles by an integral factor. There are two basic types of dividers – analogue and digital (D,4).

**DIVA** Data Input-Voice Answerback. A system of communications where a user inputs, via a data terminal, to a computer and receives a voice answerback output (O,2) of a synthesized or recorded voice.

**DJNR** Dow Jones News/Retrieval. Up-dated daily this is a computerized information retrieval system covering business news and articles in the USA from leading financial periodicals.

**DL** (1) Data Link. Equipment that permits the transmission of information between two or more data stations.
(2) Delay Line. A transmission line, circuit, device or equivalent network that has been designed to introduce a known delay (DEL,3(c)) in the transmission of a signal.

**DLC** Data Link Control. A discipline or protocol comprising a set of rules used by computers or terminals when communicating, in order to ensure the orderly transfer of information.

**DLD** Dark-Line Defect. Luminescent areas that cross an active area during lasing and are responsible for gradual laser degradation.

**DLE** Data Link Escape. The withdrawal of, or disestablishment of, a communication circuit that is used for the transmission of data in a coded form, in conjunction with data processing equipment (DPE).

**DLEC** Data Link Escape Character. The use of a control character combined with one or more succeeding characters to form an escape sequence and the development of extra data communications control operations.

**DLL** Dial Long Lines. In the USA, a range of equipment used to enable satisfactory signalling, supervision and/or transmission to be obtained on a subscriber's line that has a loop resistance higher than normal in the exchange equipment.

**DM** Delta Modulation. A form of modulation in which an

analogue signal is sampled, and the samples are quantised and converted by coding into a digital (D,4) signal. The value of a sample of the signal is compared with an integrated value of the previous sample, and a positive binary pulse is transmitted if the analogue sample is greater than the reconstructed waveform or a negative pulse if it is smaller (*also see* DPCM).

**DMA**   Direct Memory Access. A technique employed to provide the high-speed transmission of data between a computer peripheral and the computer memory without having to go through the central processor unit (CPU). Data is exchanged at maximum memory speed and several access methods to the memory are possible.

**DMAC**   Direct Memory Access Controller. A circuit or device, now available on a single chip, providing control over applications that require data to be transferred into a computer memory rather than going through the central processor unit (CPU). The DMAC can improve performance in larger systems with a number of peripherals, by completely controlling block transfers between input/output (I/O) circuits and the system memory.

**DMC**   Digital MicroCircuit. A miniaturized electronic circuit fabricated with inseparable interconnections and elements. It has been designed to accept at its input terminals particular logic states. These are converted according to set logic equations or function tables, and passed to the output (O,2) terminals.

**DMCL**   Device Media Control Language. Software used in a computer for specifying the physical use of a data base logical data structure.

**DME**   Digital Multiplex Equipment. This piece of equipment contains a digital multiplexor (MUX,2) and a digital demultiplexor (DEMUX,2). Also known as 'muldex' or 'muldem'.

**DML**   Data Manipulation Language. Not a complete computer language in itself, DML employs a host language which it then uses as a framework. It is used by a programmer to communicate between an application program and a data base.

**DMM**   Digital MultiMeter. A digital multimeter essentially converts an analogue, or continuous, signal into a series of digital (D,4) representations of sections of that signal. It ensures that the resultant display (D,6) is an accurate assessment of the signal amplitude. The DMM with its wider range of functions has largely superseded the DVM (digital voltmeter).

**DMOS** or **D-MOS**   Double-diffused Metal-Oxide Semiconductor. A short-channel device that is produced by a double-diffused

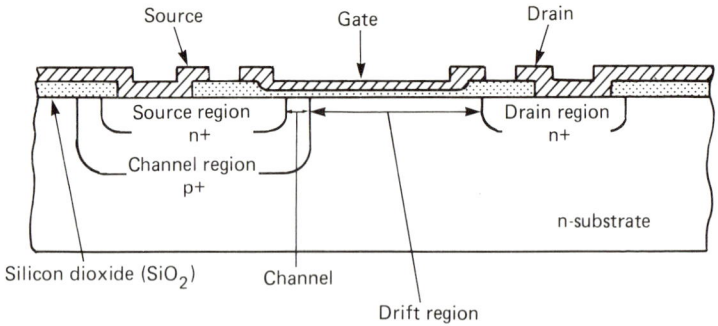

*Figure D.5* Cross-section of a double-diffused metal-oxide semiconductor (DMOS) transistor with a channel formed by double diffusion in the source region

process in which the length of the channel is determined by the difference in lateral diffusion of two impurity distributions, as shown in cross-section *Figure D.5*. DMOS devices exhibit excellent low-power and high-speed capabilities. *See* MOS.

**DMS**  Digital Multiplex Switching system. Providing a digital (D,4) circuit switched service for voice and data transmission, the DMS family of devices make use of pulse code modulation (PCM) and time division multiplexing (TDM).

**DN**  Decimal Numbering system. A system based on 10 or the powers of 10 using the digits (DGT,1) 0–9 to represent numerical quantities.

**DNA**  Digital Network Architecture. A particular design of construction for distributed computer systems developed by the Digital Equipment Corporation, who implement this architecture under the name DECNET.

**DNC**  Direct Numerical Control. The system created by a number of numerical control (N/C) machines being connected to a common memory. The memory provides data, from stored machine programs, on demand to the NC machines.

**DO**  Digital Output. An output (O,2) signal, representing the magnitude of an input signal, that is produced in the form of a series of discrete quantities coded to represent digits (DGT,1).

**DOD**  Direct Outward Dialling. The condition that allows CENTREX or PABX extensions to dial an external number without requiring the assistance of the particular exchange's operator.

**DOS**  Disk-Operating System. An organized collection of techniques and procedures for operating a computer, as software (programs or routines), held on a disk.

60

**DP** (1) Data Processing. A general term describing the organization of data in a desired manner by automatic or semi-automatic means. Also sometimes a generic term for computing in general.

(2) Distribution Point. The final point at which the conductors forming an individual telephone subscriber's circuit in a local telephone line network are run to the subscriber's premises. Known in the USA as 'ready access terminal'.

**DPC** (1) Data Processing Centre. A collection of automatic data-processing (DP, 1) equipment systems along with operators, in one centralized area under a single management group.

(2) Direct Program Control. An input/output (I/O) operation that allows direct control of a device by software programs.

**DPCM** Differential Pulse Code Modulation. A form of the process in which an analogue signal is sampled, and the samples quantized and converted by coding into a digital (D,4) signal. The actual value of every sample and its predicted value are quantized and converted by encoding into a digital signal.

**DPDT** Double-Pole Double-Throw. A particular construction of switch which can be incorporated into a double line or pole. It has two centre or common terminals each having a moveable blade connected so that each simultaneously makes contact, when activated (thrown) in one direction or the other, with its respective terminal of one of the two pairs of terminals available. *See Figure D.6.*

*Figure D.6* Symbol for a double-pole double-throw switch (DPDT)

**DPE** Data Processing Equipment. Systems comprised of inter-connected components and sub-assemblies used for data processing (DP,1).

**DPI** Device Programmer Interface. An interface between a computer and a programmer, downloading compiled program tables in either Joint Electron Device Engineering Council (JEDEC) or high level (HL) format, or uploading the tables of existing devices for editing or simulation.

**DPM** (1) Data Processing Machine. Generally refers to a machine that can store and process numeric and alphabetic information.

(2) Digital Panel Meter. A meter that provides in digital (D,4) form a display of any particular quantity or event that has been measured or recorded. Two preferred types are light-emitting diode (LED) displays for their self-illumination and contrast, and liquid crystal displays (LCD) for their low power consumption.

(3) Documents Per Minute. A measure of the rate at which media with information recorded, normally paper with printing, can be processed by equipment such as data processing equipment (DPE).

**DPS** (1) Data Processing System. A number of components and sub-assemblies connected together to receive, transmit, or store data. Many systems can also process information to a set protocol of mathematical formulae and produce desired results.

(2) Data Processing Standards. Written guide lines and procedures laid down for data processing (DP) as well as for other aspects of computing. A number of British Standards Institution (BSI) committees were concerned with these standards, which have now been brought into line with corresponding International Standards Organisation (ISO) standards.

**DPST** Double-Pole Single-Throw. A particular construction of switch which can be incorporated into a double line or pole. It has two terminals each having a moveable blade. The blades are connected so that each simultaneously makes contact, when activated (thrown), with its respective line terminal. *See Figure D.7.*

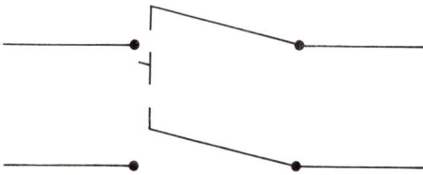

*Figure D.7* Symbol for a double-pole single-throw switch (DPST)

**DRAM** Dynamic Random-Access Memory. Called 'dynamic' because information is stored only temporarily, the dynamic random-access read/write memory (RAM) must be continually rewritten or refreshed. Storage is made in the form of the parasitic leakages that are experienced in integrated circuits

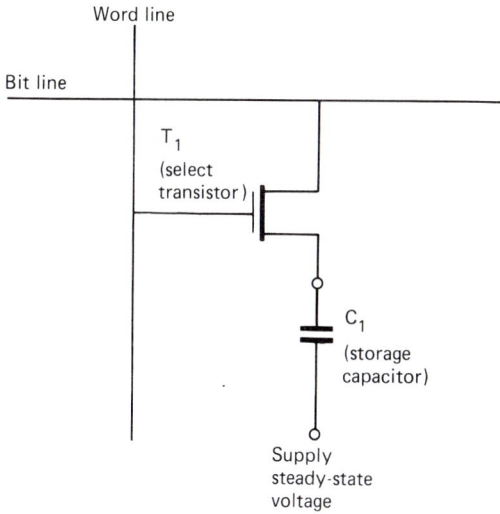

*Figure D.8* A basic dynamic random-access read-write memory (DRAM) cell

(ICs). Although the transitory nature of its storage is a drawback, the saving grace of a DRAM is its small size. *See Figure D.8.*

**DRAW** Direct Read After Write. In an optical digital (D,4) disk recording system, information once written cannot be removed. DRAW permits the recognition of errors enabling them to be corrected by rewriting data in a new position on the disk, erasing the location of the incorrect information from the memory of the computer.

**DRC** Design Rules Checking. The checking of design data for user-selected types of errors and manufacturing tolerance violations. Used in computer-aided design/computer-aided manufacture (CAD/CAM).

**DRL** Data-Retrieval Language. A computer language used in an information retrieval system, whereby information can be searched for particular items, in response to the definition of particular user needs.

**DRO** Destructive Read-Out. The phenomenon of data being lost when it is read from a storage device, all recording of it being destroyed.

**DRS** Document-Retrieval System. An information-retrieval system in which the contents of a document are indexed by means of key-word-in-context (KWIC) terms.

**DS** Data Set. A compilation of similar and related data records that have been recorded for use by a computer.

**DSA** Data Set Adaptor. Used to interface a computer to a modulator demodulator (MODEM). It reduces groups of binary digits (BITs) into individual bits for serial transmission and reverses the process for the reception of signals.

**DSB** Double SideBand transmission. A form of amplitude modulated (AM) transmission used in sound broadcasting in which frequency bands are produced above and below the carrier wave. The upper sideband is the frequency band occupied by the side frequencies above the carrier frequency. Similarly the lower sideband is the frequency band occupied by the side frequencies below the carrier frequency.

**DSC** District Switching Centre. In a four-wire-switched network, that enables any two group switching centres to be interconnected by not more than five transit links, this is the designation given to a switching centre connected to a number of group switching centres (GSC) and at least one main switching centre (MSC).

**DSCB** Data Set Control Block. A storage location, containing information necessary for the control of one data set (DS), that is needed to describe and manipulate the data set in a direct-access (DA,2) device.

**DSE** Data-Switching Exchange. A particular section of a computer network, which allows the transmission of discrete packages of information. It performs switching and routing functions and normally consists of a mini- or a microcomputer.

**DSECT** Dummy SECTion. Part of a computer program that translates a source program written in assembly language into object code. It contains data declarations but does not reserve space for that data.

**DSI** Digital Speech Interpolation. A technique to increase the capacity of long-distance digital (D,4) systems used for the transmission of speech. When a person is silent no binary digits (BITs) are transmitted; when the speech begins again, bits flow. It will be used in INTELSAT satellite communications.

**DSL** Data Set Label. The identification given to a storage location containing information necessary for the control of one data set (DS).

**DSR** Data-Set Ready. A line on a modulator demodulator (MODEM) which informs the transmitting terminal that the signal received is all right. An RS-232 standard (*see* RS-232C).

**DSV** Digital Sum Variation. An expression of the difference between the maximum and minimum values of the digital (D,4) sum of a coded signal.

**DSW** Device Status Word. A computer word that contains binary digits (BITs), the condition of which indicates the particular status of various devices.

**DT** Data Transmission. (a) The sending of data by means of signals sent over a telecommunication network, from a data source to a data sink. (b) The sending of data from one part of a system to another.

**DTE** Data Terminal Equipment. (a) A unit at a data station that can receive transmitted data signals and/or transmit data. It can also control information transfer according to the protocol of the data link network. (b) Any piece of equipment at the beginning or end of a communications path. It can be a terminal or computer or the end user of a telephone circuit.

**DTL** Diode-Transistor Logic. A combination of diodes and transistors in an integrated circuit (IC) form that comprises the logic elements of a system. A basic DTL logic gate design is shown in *Figure D.9*.

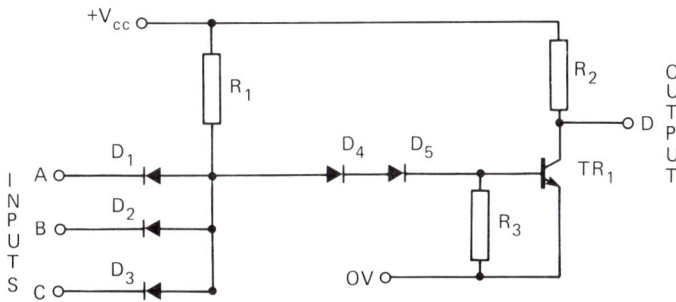

*Figure D.9* A basic diode-transistor logic (DTL) gate design

**DTMF** Dual-Tone MultiFrequency. A name given to telephone pulsing where every digit is represented by a particular pair of audio frequencies, one below $1000\,Hz$, the other above $1200\,Hz$.

**DTR** Data Terminal Ready. A line on a terminal which informs the modulator/demodulator (MODEM) that it is ready to transmit data.

**DTS** Data-Transmission System. A series of devices, circuits or modulator/demodulators (MODEMs), that can translate or transfer information from one position to another.

**DTSS** Dartmouth Time Sharing System. An interactive computer system developed at Dartmouth College in the USA which was operated in 1969. The system enables non-computer science specialists to use the system, as it provides user-friendly facilities.

**D-type** A connector employed in both the data processing (DP,1) and telecommunication (TELECOMM) industries. Available in sizes from 9 to 37 ways, it is usually used for interconnecting between equipment or for connecting different assemblies within one unit. The name derives from the elongated D shape of the pin housing.

**DURAL** DURALumin. An alloy of aluminium containing copper, manganese and silicon and having a high-tensile strength. It has many uses where conduction and movement are required, e.g. switchgear moving parts.

**DUV** Data Under Voice. A system of digital (D,4) data transmission carrying information on a frequency below that used for voice transmission. *See* DAV (2).

**DVM** Digital VoltMeter. A meter that displays (D,6) in a digital (D,4) form, measurements of alternating current (AC) or direct current (DC,1) voltages. It has generally been superseded by the digital multimeter (DMM).

**DWBA** Direct-Wire Burglar Alarm. A wired electrical circuit used for the transmission of alarm signals from a protected premises to a monitoring station, often a police station.

**DWM** Destination Warning Marker. A warning spot placed on magnetic tape (MAGTAPE) about 2 metres from its end. The spot is reflective and can be sensed photoelectrically indicating the proximity of the end of the tape.

**DX** (1) DupleX. A term describing the simultaneous transmission over a channel (CHNL(a)), link or circuit in both directions.

(2) DupleX signalling. A signalling system which occupies the same cable pair as the voice path, but does not require any special type of filtering.

(3) Long distance communication. The success achieved when DXing (long-distance radio listening and often two-way communications) depends substantially on prevailing propagation conditions, which include local weather conditions, various atmospheric conditions and sunspots.

(4) Distance. The number of corresponding digit positions which differ between two binary words of the same length. The distance of the two binary words 11001101 and 10110011 is 6.

# E

**E** (1) Exa. A decimal prefix used in association with a base unit in the SI system of units indicating a multiplication factor of $10^{18}$

of that unit. For example, one exatonne, Et, is equal to
1 000 000 000 000 000 000 tonnes.

(2) Erlang. A unit expressing intensity of traffic. One
continuously-engaged line has a traffic flow of one erlang; if the
average number of calls made at any one time is 12, then the
traffic intensity is 12 erlangs.

(3) The hexadecimal (HEX) symbol for binary 1110, decimal
14.
$E_{16} = 14_{10} = 16_8 = 1110_2$.

**EA** Effective Address. The location or unit in a computer, where
information is stored that is actually used for the execution of
an instruction. This may differ from that of the instruction in
storage.

**EAE** Extended Arithmetic Element. A central processor element in a microprocessor that is implemented by hardware to
multiply, divide and normalize functions.

**EAM** Electrical Accounting Machine. A term that can be
applied to any one of the many electro-mechanical devices
employed in data processing (DP,1). It can include tabulators
and calculators but excludes digital (D,4) electronic devices.

**EAN** European Article Number. A computer coded number in
the form of a bar code. Bar codes are used on consumer
products as identification.

**EAPD** ElectroAbsorption PhotoDiode. An avalanche photodiode made from a direct-gap semiconductor material, like
gallium assenide (GaAs), having a low carrier concentration.
The responsivity and quantum efficiency are high at
wavelengths close to and beyond the absorption edge in GaAs.
The avalanche gain at these wavelengths is much higher than
the gain at shorter wavelengths.

**EAPROM** Electrically Alterable Programmable Read-Only
Memory. A specialized read only memory (ROM). The
contents can be erased en masse in one operation by applying a
higher than normal voltage to a programme input. The
information held in the memory can be altered completely or
partially, some special types allowing a single word to be erased
and rewritten. These nonvolatile memories have no fusible
links (FL) and do not require ultra-violet (UV) irradiation.

**EAROM** Electrically-Alterable Read-Only Memory. A specialized random-access read/write memory that can be programmed by writing into the array and then generally used as a read
only memory (ROM). The devices have a fast read cycle of
around 15 microseconds and a much slower write cycle of about
one millisecond. Other qualities are as EAPROMs (*see above*).

**EAX** Electronic Automatic eXchange. A telephone exchange

employing modern techniques such as digital (D,4) switching, in which communication between subscribers is effected without the intervention of exchange operators.

**E/B** Electrode per Bit. A form of operation of a charge-coupled device (CCD) memory that increases the number of binary digits (BITs) of information that can be stored.

**EBCDIC** Expanded (or Extended) Binary Coded Decimal Interchange Code. An 8-binary digit (BIT) code, similar to American Standard Code for Information Interchange (ASCII), that uses the eighth level as an information bit. The code is used to represent 256 unique letters, numbers and special characters.

**EBR** Electron Beam Recording. A process that employs an electron beam to store and read information on a particular target. These targets are normally silicon dioxide and electrostatic charges actually store the data. On material such as vitrium iron garnet the data is stored as magnetic bubbles.

**EBU** European Broadcasting Union. An organization of radio and television broadcasting authorities from a number of European countries.

**EC** Error Correcting. The process of detecting invalid characters that will not conform to a set of rules. The construction of the rules is held in an error-correcting code.

**ECAP** Electronic Circuit Analysis Program. A computer language used for the modelling and analysis of electrical networks. It employs function generators or tables of functions along with passive components for close examination of device models.

**ECB** Event Control Block. A group of words or records treated as an individual unit that indicate the status of predictable events and take appropriate action where necessary.

**ECC** Electronically-Controllable Coupler. In fibre optics (FO), this is part of an optical system which enables other parts to be coupled or uncoupled from each other via an electrical signal.

**ECCM** Electronic Counter-CounterMeasures. Electronic equipment, devices and techniques employed to destroy or degrade the effectiveness of an enemy's electronic countermeasures (ECM).

**ECD** ElectroChromeric Display. Low-voltage, low-current devices that are chemically transformed from transparent to opaque when excited by an external electric field. When the field is turned off the display (D,6) continues to be opaque until subjected to a field with a polarity opposite to the first. The displays created are legible over a wide viewing angle and are compatible with integrated circuit (IC) technology.

**ECG** ElectroCardioGraph. An instrument which produces a trace (called an 'electrocardiogram') that provides a record of the voltage (V) and current (I) emanating from the hearts of living animals or humans.

**ECH** Eddy-Current Heating. Heat produced in a conductive material, induced by the material being subjected to a varying magnetic field (eddy-current) usually from a coil with an alternating current (AC).

**ECHO** European Commission Host Organization. An organization that makes available a computerized referral and enquiry service, as well as providing information to a number of stores of computer files.

**ECL** Emitter-Coupled Logic. A family of circuits that have been designed to perform a specific logical function. Pairs of transistors that form the logic circuit are connected together by their emitters, as shown in *Figure E.1*.

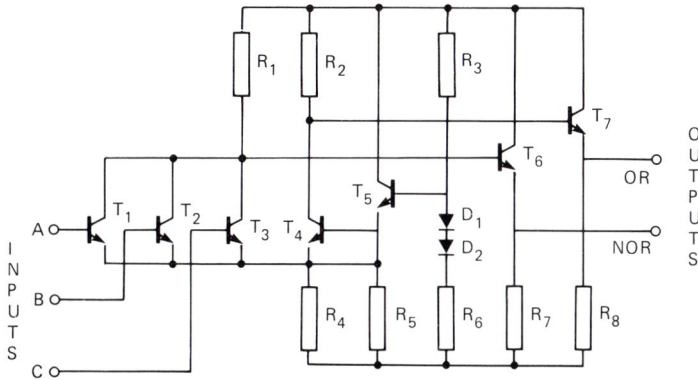

*Figure E.1* An emitter-coupled logic (ECL) OR/NOR gate circuit

**ECM** Electronic CounterMeasures. Electronic jamming equipment and other devices and techniques used in tactical and strategic applications to destroy the effectiveness of an enemy's electronic aids to warfare, as a function of electronic warfare (EW).

**ECMA** European Computer Manufacturing Association. An organization comprising original equipment manufacturers (OEM) of computers from a number of European countries.

**E-COM** EleCtrOnic Mail. A service for the electronic transmission or distribution of messages, provided by the US postal service.

**ECS** (1) European Communications Satellite. Intending to provide Europe with a satellite communications system, this programme is run by the European Space Agency (ESA).
(2) Experimental Communication Satellite. A range of experimental communication satellites put up by Japan.

**EDA** Electronic Differential Analyser. An analogue computer, using a number of interconnected integrators, employed in solving differential equations.

**EDP** Electronic Data Processing. The conversion of data by electronic equipment into a form able to be processed by computers. EDP is a term used more often in the USA whilst data processing (DP) is more often used in the UK and Europe.

**EDPM** Electronic Data-Processing Machine. An individual unit or device that either is part of an electronic data processing system (EDPS), or can provide stand-alone electronic data processing (EDP).

**EDPS** Electronic Data-Processing System. A collection of electronic data processing machines (EPDMs) that are capable of performing electronic data processing (EDP) functions.

**EDS** Exchangeable Disk Store. A generic term used to describe storage disks that are capable of being removed from a computer and raplaced by others.

**EDSAC** Electronic Delay Storage Automatic Computer. Developed in 1949 by the mathematics department of the University of Cambridge, EDSAC was one of the first electronic computers.

**EDVAC** Electronic Discrete Variable Automatic Computer. One of the first electronic computers, designed and built in 1949 at the University of Pennsylvania.

**EEC** European Economic Community. Commonly known as the 'Common Market', the EEC was founded by the treaty of Rome in 1957, and was extended in 1973. Its membership includes Belgium, Denmark, France, West Germany, Greece, Ireland, Italy, Luxembourg, Netherlands and the United Kingdom.

**EEG** ElectroEncephaloGraph. An instrument which produces a trace (called an 'electroencephalogram') that provides a record of the voltage (V,2) waveforms which emanate from the brains of animals and humans.

**EEL** Exclusive Exchange Line. Also known as a direct exchange line, this is a single telephone line that provides service to an individual subscriber and has only one exchange number.

**EEPROM** Electrically-Erasable Programmable Read-Only Memory. Using only silicon and its derivatives plus metals, this device employs the floating-gate tunnel oxide (FLOTOX) cell structure in which electrons tunnel to and from a floating gate. *See Figure E.2.* It is a field programmable read-only memory (FROM) in which the cells may be erased electrically in one second and may be reprogrammed electrically. It can also be erased and reprogrammed up to a million times. This memory device in general makes practical such applications as altering remotely, by telephone, microprocessor system programs. Also known as $E^2$PROM.

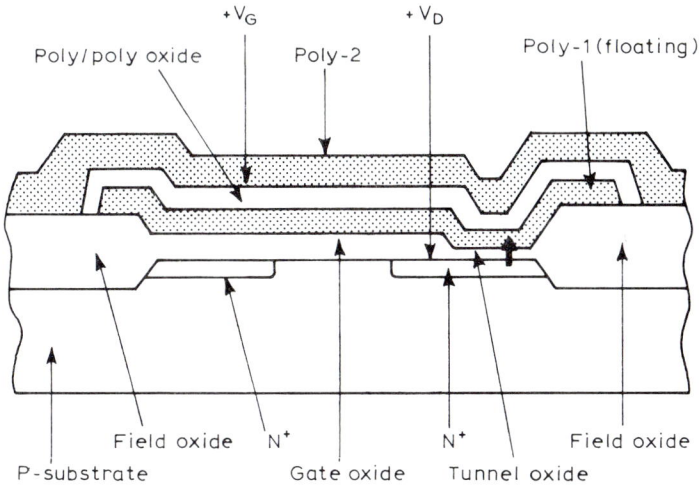

*Figure E.2* A cross-section of an electrically-erasable programmable read-only memory (EEPROM) that employs the floating-gate tunnel

**EET** Equipment Engaged Tone. A tone heard by a caller if any part of a telephone network is overloaded and the call cannot be completed. An engaged tone (ET) or busy tone (BT,2) is used for this purpose in many systems replacing an EET, although the busy tone actually indicates that a called party is busy.

**EFL** Emitter Follower Logic. A type of bipolar integrated circuit (IC) fabrication technology used in logic circuits.

**EFTA** European Free Trade Association. A collection of trading countries within Europe which includes Austria, Norway, Portugal, Sweden, Switzerland and Iceland, with Finland as an associate member.

**EFTS** Electronic Funds Transfer System. A general term used in the USA that refers to various electronic communications systems that are able to transfer financial information from one point to another. The term can encompass automated clearing houses and point-of-sale (POS) systems.

**EHF** Extremely-High Frequency. Frequencies occurring between 30 GHz and 300 GHz, forming part of a larger continuous range of frequencies. EHFs are considered by international agreement to be radio frequencies in band number 11 and have wavelengths of between 1 mm and 1 cm.

**EHOG** European Host Operators Group. An organization representing the interests of a number of information vendors, who hold various stores of computer files, within Europe.

**EHT** Extra-High Tension. (a) Generally a term used to describe the high-voltage supply for display (D,6) tubes. The displays can be either cathode-ray tubes (CRTs) or television pictures tubes. (b) The term is also part of power distribution terminology.

**EHTPS** Extra-High-Tension Power Supply. A form of power-supply unit (PSU) that can provide a high voltage (HV,2), normally several thousand volts (kV) output (O,2), from a lower voltage input. *See* EHT.

**EIA** Electronics Industries Association. A trade association of the electronics industry in the USA. It formulates technical standards, such as the widely used RS-232C interface standard for connecting computer equipment, and also maintains contact with American government agencies in matters relating to the electronics industry. It is one of the bodies that help determine standards for the USA Standards Institute (USASI).

**EIES** Electronic Information Exchange System. Set up in the USA, this experimental computer network is supported by the National Science Foundation. Connecting a number of groups together, it provides electronic mail (E-COM), teleconferencing and personal electronic communications, whilst monitoring their use and impact.

**EIN** European Informatics Network. A heterogeneous packet-switched network set up by agreement between a number of European countries.

**EIRP** Effective Isotropically Radiated Power. A comparison made to a given reference antenna, by the relative gain in a given direction for a particular power supplied to an antenna. Known in the USA as 'effective radiated power' (ERP).

**ELCB** Earth-Leakage Circuit-Breaker or Earth-Leakage Contact-Breaker. A device that provides a disconnection or break in electrical supply if the voltage in the earth or non-current-carrying part of a circuit, or the out-of-balance current in the supply due to leakage, exceeds a predetermined level.

**ELD** Electro Luminescent Display. An illuminating device used to backlight a liquid crystal display (LCD), so that it can be used in a darkened environment. The unit is contained in a plastics sleeve for protection from moisture, is available in a number of low-reflection colours providing uniform brightness and has a long dependable life.

**ELF**   Extremely-Low Frequency. A frequency less than 100 hertz (Hz).

**ELINT**   ELectronic INTelligence. Electronic techniques employed in signal analysis and location as tools for electronic warfare (EW). These are combined with the interpretation of the information gained concerning an enemy's activities.

**ELSI**   Extra-Large-Scale Integration. The design and production of integrated circuit (IC) or microcircuit chips (CHIP) in which the solid state circuitry is manufactured on a single piece of material, normally silicon. ELSI refers to circuits with a capability of more than one megabit. Sometimes also referred to colloquially as 'Even-Larger-Scale Integration'.

**ELV**   Extra-Low Voltage. In European regulations a term used to imply a voltage not exceeding 50 volts (V).

**EM**   End of Medium. A specific control character used to indicate the physical end of a data medium. It normally allows sufficient time for the user to adjust or change the system of operation.

**EMC**   ElectroMagnetic Compatibility. The ability of generators of electromagnetic radiation and equipment susceptible to interference from such radiation, to exist in the same environment without degradation of performance. This is achieved generally by reducing the level of radiated electromagnetic interference (EMI), by introducing attenuation in all EMI paths between generator and susceptible equipment.

**EMF**   ElectroMotive Force. The physical property of a device that enables it to drive an electric current around a circuit. The property can be produced by either mechanical or chemical means, for example a generator or a battery.

**EMI**   ElectroMagnetic Interference. The distortion or impairment of a desired transmission or signal brought about by an electromagnetic disturbance.

**EMI/RFI**   ElectroMagnetic Interference and Radio Frequency Interference. A range of conducted or radiated electromagnetic emissions that produce an unintentional and unwanted response in a piece of electronic equipment or a computing device.

**EMIS**   Electronic Materials Information Services. A store of information held on computer files concerning material properties and material supply. It is produced by the Information Services: Physics Electrical and Electronics, and Computers and Control (INSPEC) body.

**EMMS** Electronic Mail and Message System. The services in a communications system or network such as electronic mail, teleconferencing, word processor communication and videotex.

**EMP** ElectroMagnetic Pulse. High-energy radiation, such as that produced by a nuclear blast.

**EMS** Electronic Message System. First used as a description of communication made using terminals in a communications network or system. It is now generally employed to describe specialized services such as electronic mail, teleconferencing, word processor communication and videotex.

**EMW** ElectroMagnetic Wave. The resultant effect obtained from the interaction of an electric field and magnetic field. The interaction causes energy to be transmitted in a particular direction.

**END** A computer statement used to denote the end of a computer program.

**ENIAC** Electronic Numerical Integrator And Calculator. Considered by many to be the first truly electronic computer, ENIAC was produced by the University of Pennsylvania during the Second World War. It contained more than 18 000 electron tubes.

**ENQ** ENQuiry. A special control character used in data communications. It has been designed to elicit a response from a remote station or terminal. The response usually required is some form of identification and/or operating status.

**ENR** Excess Noise Ratio. A ratio of the noise output of a particular device against a noise temperature of 290° kelvin (K,2) (which is a defacto standard reference point).

ENR (in dB) = $10 \times 10g10$ (TE/TO $- 1$)

TE = Noise temperature of device

TO = 290°K.

**EO** End Office. Class 5 of five class numbers assigned to offices according to their function, used in the American direct distance dialling (DDD) telephone network. Any one centre can handle telephone call traffic from one, two or more centres lower in the hierarchy, Class 1 being the top classification.

**E-O** ElectroOptic. Relates to the conversion of electrical power to optical power, (electrical signal to optical signal conversion).

**EOA** End Of Address. A specific control character used to indicate the finish of characters that are not text, i.e. they are address information, employed on a transmission line.

**EOB** End Of Block. A control character used to indicate the finish of a group of words or records treated as an individual unit, employed on a transmission line.

**EOF** End Of File. (a) A mark used at the end of a quantity of

74

data to indicate its completion. (b) A computer statement that indicates the end of a file of information.

**EOJ** End Of Job. (a) A status condition that exists in a system when a particular job has been completed. (b) A computer statement that indicates the end of a task or job.

**EOL** End Of Line. A specific character that indicates the termination or finish of a group of records.

**EOM** End Of Message. A specific character or set of characters that indicate the termination or finish of a message or record.

**EOR** (1) End Of Reel. A character or word employed to indicate the termination or finish, in a multi-reel magnetic tape file, of all but the last reel.

(2) End Of Record. A computer statement used to indicate the finish of a record or a set of records.

(3) End Of Run. A computer statement used to indicate the finish of a computer run.

**EOT** (1) End Of Transmission. An indicator consisting of a unique group of characters at the termination of a transmission. It denotes the end of a data transmission to or from a remote terminal and also resets stations in the transmission line to control mode.

(2) End Of Tape. A special mark that can be either a specific code on the magnetic tape (MAGTAPE) or a physical indicator used to show the physical end of a reel of tape. The physical indicators can be in the form of an easy-to-see reflective strip or a transparent section.

**EOV** End Of Volume. (a) A marker that indicates the finish of a physical volume. (b) A computer software routine that comes into operation and takes control when the situation in EOV (a) is indicated.

**EPB** Environmental Periodicals Bibliography. A store of information held on computer files, in the USA, that relates to aspects of environmental sciences.

**EPIC** Exchange Price IndiCation. A store of information held on computer files, managed by the London Stock Exchange. It contains information from a number of international sources.

**EPROM** Erasable Programmable Read-Only Memory. A read-only memory (ROM) that can have information that is stored in its memory changed over and over again. *See Figure E.3.* EPROMs that use metal-oxide semiconductor (MOS) technology can be erased by ultra-violet (UV) light through a small window in the top of their packaging.

**E$^2$PROM** Electrically-Erasable Programmable Read-Only Memory. Using only silicon and its derivatives plus metals, this device employs the floating-gate tunnel oxide (FLOTOX) cell

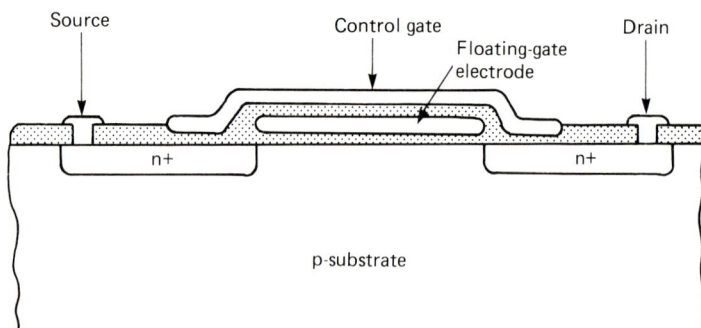

*Figure E.3* A cross-sectional view of an erasable programmable read-only memory (EPROM)

structure in which electrons tunnel to and from a floating gate. *See Figure E.2.* It is a field programmable read-only memory (FROM) in which the cells may be erased electrically and may be re-programmed electrically. Also known as EEPROM.

**EPSS** Experimental Packet-Switched System. An experimental packet-switched computer network set up by the Post Office in the UK in 1979.

**EQ** EQual to. A comparison of two values that states whether a second value is equal to the first. The symbol is =.

**ER** Error Recovery. A system or procedure that has been designed to overcome the deviation of a computed or a measured quantity from the theoretically correct or true value.

**ERP** Effective Radiated Power (US). A comparison made to a given reference antenna, by the relative gain in a given direction for a particular power supplied to an antenna. Known in the UK as 'effective isotropically radiated power' (EIRP).

**ESA** European Space Agency. Responsible for the cooperative development of launch vehicles and satellites for scientific, weather and communication use, ESA was founded in 1975.

**ESA-IRS** European Space Agency-Information Retrieval System. With its headquarters in Italy, the European Space Agency (ESA) provides access to a number of stores of information held on computer files, relating to scientific matters.

**ESC** (1) EScape Character. A special symbol, part of a character set, employed to show that following characters belong to a different character set from those that have gone before.
(2) Edison Screw Cap. A lamp cap or bulb cap, with the outer case in the form of a screw thread which provides one contact, the other contact being a central end projection.

**ESI** Externally Specified Index. Enabling a computer system to become a central message switching centre, for a variety of remote devices, this facility automatically routes messages to and from a main store. It achieves this without the programme sequence of the central processor unit (CPU) being disturbed.

**ESR** Equivalent Series Resistance. The amount of resistance in series with an ideal or loss-less capacitor (C,6) which exactly duplicates the performance of a real capacitor. In general, the lower the ESR, the better the quality of the capacitor and the more effective it is as a filtering device. It is a prime determinant of ripple in switching power supplies.

**ESS** Electronic Switching System. Originally a term for a Bell System computerized telephone exchange, ESS is now used generally as an industry term.

**ET** Engaged Tone. In a telephone system that employs an equipment engaged tone (EET), an ET or busy tone (BT,2) indicates to a caller that the called party is engaged. In systems not employing EET, an ET (as in the UK) can indicate that any part of a telephone network is busy.

**ETB** End of Transmission Block. A control character employed to indicate the termination or finish of transmission of a group of words or records treated as an individual unit.

**ETIM** Elapsed TIMe. The period of time measured from the start to the completion of a process. This apparent time period can sometimes be longer than the time actually taken in processing information.

**ETV** Educational TeleVision. Available to television receivers in the home, this form of broadcasting informs and teaches. UK examples are schools' programmes and the Open University.

**ETX** End of TeXt. A specific control character employed to indicate the end of transmission of a piece of text.

**EURODICAUTOM** EUROpean AUTOMatic DICtionary. A machine aided translation (MAT) system employing a bank of specific terminology that can assist in the translation of scientific and technical documents.

**eV** Electron-Volt. A unit of energy that can be defined as the amount of energy released or acquired when an electric charge or one electron moves through a potential difference (PD) of one volt (V).

**EVA** Ethylene Vinyl Acetate. A material used to improve the handling capabilities of optical fibres (OF) protected by a thin (5 to $10\,\mu$m) plastics coating. The improvement is affected by the addition of a second, much thicker, coating ($250\,\mu$m) of EVA.

**EW** Electronic Warfare. The use of electronic devices for the purposes of: determining the existence and disposition of an enemy's electronic aids to warfare; destroying or degrading the effectiveness of an enemy's electronic aids to warfare; preventing the destruction of the effectiveness of friendly electronic aids.

**EVCB**  EVent Control Block. An area of storage used by a computer operating system (OS) to record the current condition or status or an event.

**EVR**  Electronic Video Recording. The recording of video signals electronically on magnetic tape, magnetic disk or any other suitable material.

**EXEC**  (1) EXECutive statement. A job control character that labels the load module to be executed along with specific job steps.

(2) EXECute. To carry out the operations specified by a particular instruction or routine.

**EXTRN**  EXTeRNal reference. An indication of where to find a single variable from a range, or where to find an item which is not defined in a particular computer program, segment or subroutine.

# F

**f**  Femto. A decimal prefix commonly used in association with a base unit in the SI system of units. It indicates a multiplication factor of $10^{-15}$ of that unit. For example, one femtosecond, fs, is equal to 0.000 000 000 000 001 second.

**F**  (1) The hexadecimal (HEX) symbol for binary 1111, decimal 15, the highest hexadecimal digit.
$F_{16} = 15_{10} = 17_8 = 1111_2$.

(2) Farad. Unit of capacitance, defined as the capacitance (C,5) of a capacitor (C,6) which requires a potential difference (PD) of 1 volt (V) to maintain a charge of 1 coulomb (C,9) on that capacitor. Named after Michael Faraday who discovered electromagnetic induction in 1831.

(3) Final. In computing this is a subscript, or lower case, symbol that designates *final*.

**FACE**  Field-Alterable Control Element. A chip (CHIP) employed in some microcomputer systems to permit the user to write microprograms. It allows for field microprogramming in low-volume applications and each system has a writeable control store, a logic unit and a display-and-debug unit.

**FACT**  (1) Fast Access Current Text. An experimental electronic library, able to produce documentation facsimile (FAX) copies from microfiche via a telephone network.

(2) Fully Automated Cataloguing Technique. A library system that combines computer control of documents, along with storage on microfilm and a computer output microfilm (COM) facility.

**FAIRS** Fully Automatic Information Retrieval System. A retrieval system that provides automatic searching of computer files containing information, in response to the definition of a particular user's needs.

**FAMOS** Floating-gate Avalanche-injection Metal-Oxide Semiconductor. A particular type of erasable programmable read only memory (PROM) that erases by use of ultra-violet (UV) light and has an avalanche-transport mode. *See Figure F.1*. It has silicon-gate metal-oxide semiconductor (MOS) field-effect transistors (FET) that have no connection to the floating gate, and memory operation depends upon charge transport effected by avalanche injection from a source or a drain.

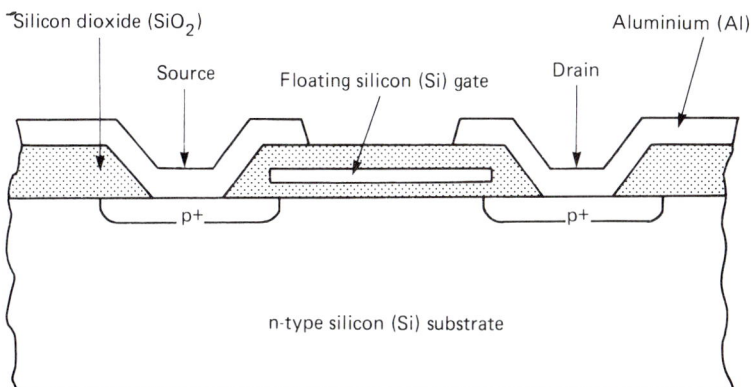

Silicon dioxide ($SiO_2$)    Aluminium (Al)

Source    Floating silicon (Si) gate    Drain

p+    p+

n-type silicon (Si) substrate

*Figure F.1* A cross-sectional view of a floating-gate avalanche-injection metal-oxide semiconductor (FAMOS) structure

**FAP** Fortran Assembly Program. A procedurally orientated formula translater (FORTRAN) computer software system.

**FAST** Flexible Algebraic Scientific Translator. Device enabling a user to write a computer program in an algebraic format by translating complex statements into Beginner's All-purpose Symbolic Instruction Code (BASIC).

**FAX** FACSimile. A system used for the transmission of printed images. During operation the image is scanned by equipment at the transmitting station, transmitted and then reconstructed by the receiving station and finally duplicated as hard copy on paper.

**FAXCOM** FASCimile COMmunications. A service offered in North America by the TransCanada Telephone System. It allows for the transmission of written, drawn or typed graphics over a direct distance dialling (DDD) network.

**FCC** Federal Communications Commission. Appointed by the President of the USA, this is a board comprising seven commissioners who have the power to regulate all interstate and foreign electrical communication systems originating in the USA.

**FD** (1) Floppy Disk. A mass-storage device employing a soft flexible (floppy) diskette coated with magnetic material onto which information is recorded. The disk(ette) is driven so that it rotates inside a cardboard jacket and cutouts are provided for a moving head and index hole information.

(2) Full Duplex. The designation given to a method of operating a communications circuit so that each of its ends can simultaneously transmit and receive.

**FDC** Floppy-Disk Controller. Controls the transfer of data to and from a floppy disk (FD,1). FDCs can interface both single-density and double-density disk drives and are able to interface up to four double-density drives. High-performance disk controllers can feature automatic seek, read and write operations, select lines for multiple heads, programmable loading and unloading of heads, control lines for phase-locked loops (PLLs) generally required for disk-drive circuitry and operate on a 5 volts (V) power supply unit (PSU).

**FDF** Fibre Distribution Frame. A type of optical fibre (OF) connector panel, whereby distribution and jumper cables, usually of a single fibre each, which can splice or connect optical transmitters to optical receivers, are able to be joined.

**FDM** Frequency-Division Multiplexing. A method of multiplexing in which more than one signal is transmitted on a channel (CHNL(a)) by dividing the available transmission frequency into narrower bands. A number of separate single channels are thus derived from splitting up a wider bandwidth into several narrower bandwidths.

**FDMA** Frequency-Division Multiple Access. A technique enabling communicating devices at various different locations to share a multipoint or broadcast channel (CHNL(a)). It allocates a different frequency to each user.

**FDOS** Floppy-Disk Operating System. (a) An operating system (OS) that has been developed to permit source programs written and edited at the system keyboard, to be stored on floppy disks (FD,1). The assembled programs are stored on the disk in binary form. A typical FDOS package can consist of dual drive floppy disk, interface, software, documentation and cable. (b) A combination of areas in the memory map (MAP,2) of a microcomputer using a control program for microcomputers (CP/M) disk operating system (DOS). The areas are the basic input/output system (BIOS) and the basic disk-operating system (BDOS).

**FDX**  Full DupleX. The property of a communication system or particular piece of equipment to communicate simultaneously in two directions (*see* FD,2).

**FE**  Format Effector. A specific control character used for the control of the layout or position (format) of printed or displayed information.

**FEDLINK**  FEDeral Library and Information NetworK. A computer network in the USA providing an on-line or user direct cataloguing facility.

**FEDREG**  FEDeral REGister. A store of information held on computer files providing details of US Federal Government regulations.

**Fe₂O₃**  Ferric Oxide. This red iron oxide is the magnetic constituent of nearly all magnetic recording tapes (MAGTAPE) used today. It is dispersed in fine particles to form a coating on the tapes.

**FEP**  Front-End Processor. The interface between a mainframe computer and the outside world. It is normally used to remove some of the necessary processing load from the computer.

**FET**  Field-Effect Transistor. A semiconductor device with unipolar multielectrodes, where current flows through a narrow semiconductor material conducting channel (CHNL(h)) between two electrodes and is modulated (made more or less conductive) by means of an electric field applied at a third electrode. *See Figure F.2.* This three-terminal device consists of the channel which is a slice on n-type silicon (n-type) with a p-n junction (p-n) diffused into it. As FETs are controlled by voltage (V) rather than current (I), they exhibit low power requirements along with high impedance.

**FEXT**  Far-End CROSSTalk. (a) Observation and measurement at a remote terminal in a telecommunications system of unwanted energy transfers between one circuit and another which allow the unwanted energy to travel in the same direction as the wanted signal. (b) In optical fibres (OF) the similar condition can be experienced as in (a). Measurement is at the receiving end of a given fibre when optical power is inserted into an adjacent fibre at its transmitting end.

**FF**  (1) Form Feed. (a) A control character that provides regulation of the demands of a printer to feed forms. (b) A

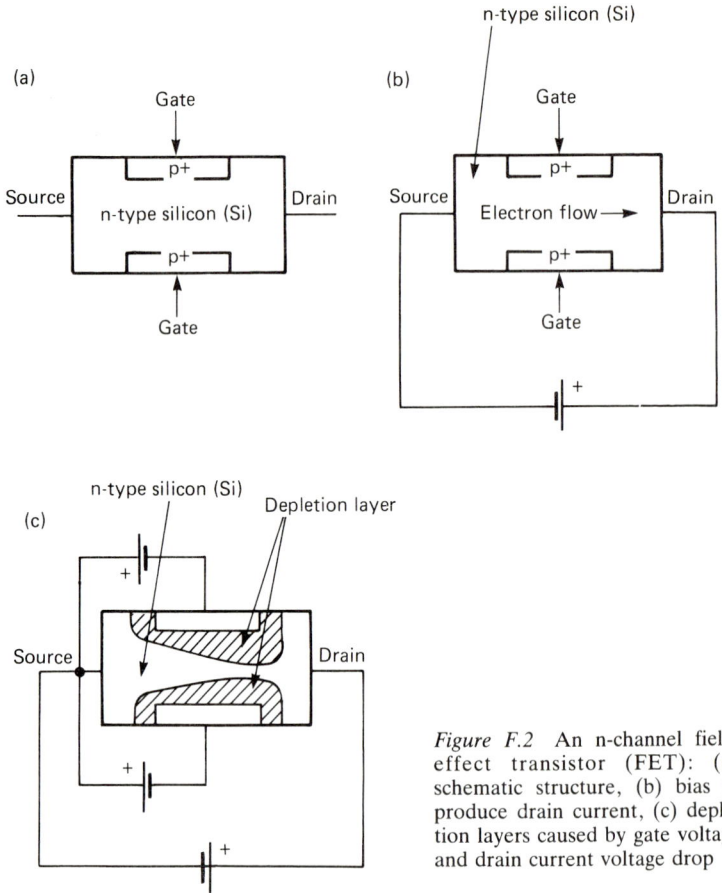

Figure F.2 An n-channel field-effect transistor (FET): (a) schematic structure, (b) bias to produce drain current, (c) depletion layers caused by gate voltage and drain current voltage drop

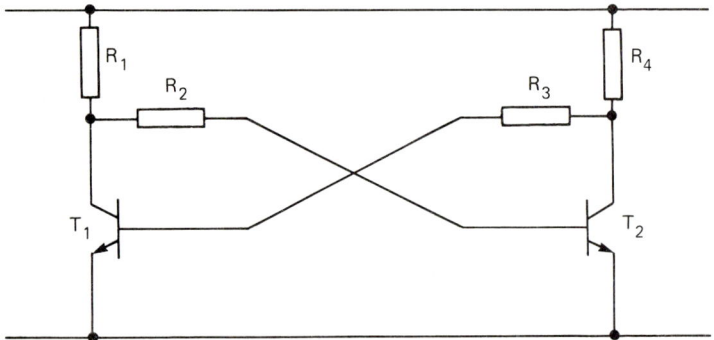

Figure F.3 A bistable circuit or flip-flop (FF,2)

mechanical form-positioning system on a printing device that automatically locates and positions continuous stationery.

(2) Flip-Flop. A combination of two inter-connected active devices (i.e. a bistable) or two inter-connected logic gates. As long as power is available an FF circuit can be used to store one binary digit (BIT) of information permanently. *See Figure F.3.*

**FFL**  Front Focal Length. A distance measured from the focal point, located in front of a lens in an optical system, to the nearest point on that lens.

**FFT**  Fast Fourier Transform. Any realistic periodic signal can be represented as the sum of sine waves having various amplitudes and phases. This sum, although an infinite series, converges quickly. To include transients or non-periodic fucntions, the method for Fourier-series analysis is extended. In such a case the infinite-series becomes a continuous function expressed as an integral, known as the 'Fourier transform'. To make available a method whereby calculations can be performed by computer, the Fourier transform (an infinite series) is provided with a finite range, known as the FFT.

**FIFO**  First In, First Out. In memories, tables, stacks or lists, this is a method of storing and retrieving data or items. Information is deposited at one end, and removed from the other, enabling the oldest entry to be retrieved first.

**FIGS**  FIGures Shift. A function initiated by a figure shift character and performed by a teletype that causes the machine to make a physical movement to the upper case in order that numbers, symbols and upper case characters can be printed.

**FILO**  First In Last Out. Also known as a 'stack', this is a pop-up register which enables the most recent entry to be retrieved first.

**FIMS**  Functionally-Identifiable Maintenance Systems. A system of block diagrams organized so that it can direct non-technical personnel to a repleaceable unit. The unit can range from a computer (CMPT) to a printed circuit board (PCB), or to a simple repairable circuit (CCT).

**FIP**  (1) Vacuum Fluorescent Indicator Panel. A specialized variation of a triode, having a directly heated cathode (filament), control grid, and segmented plate (anode) upon which a fluorescent material has been deposited. *See Figure F.4.* An electron flow from the filament is directed upon one or more of the anode segments, causing the phosphor on them to luminesce and emit light at a high brightness level. As the fluorescent material used is an efficient converter of electrical energy to light energy, FIPs operate at low voltages (V).

(2) Federal Information Processing standards. American

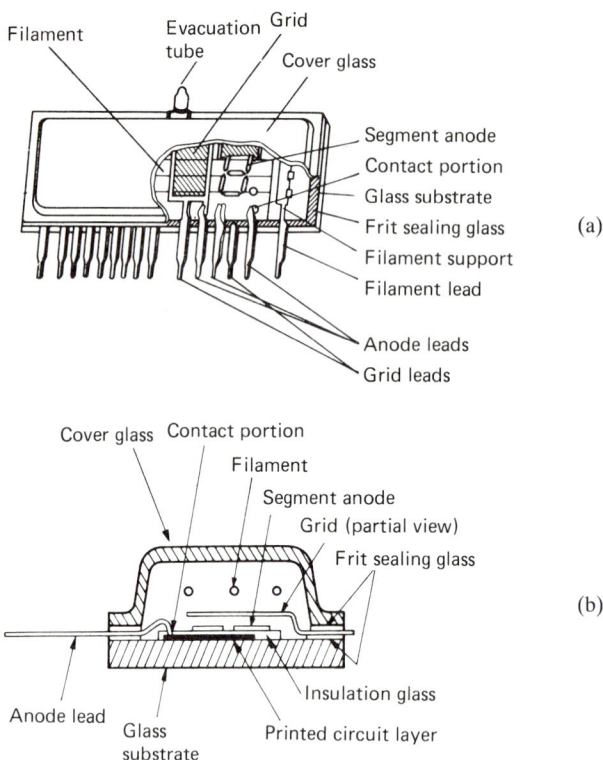

*Figure F.4* Vacuum fluorescent indicator panels (FIP): (a) construction, (b) cross-section (*courtesy* NEC Electronics (UK) Ltd)

National Standards Institute (ANSI) standards which have been adopted by the US Government.

**FIRST** Fast Interactive Retrieval System Technology. An information retrieval technique that provides a system having the ability to allow the searching of computer files that contain particular items of data. The search is implemented under real time operator control.

**FJCC** Fall Joint Computer Conference. Originally organized by the American Federation of Information Processing Societies (AFIPS), these conferences that take place in the USA have now been superseded by the National Computer Conference (NCC).

**FL** Fusible Link. A fusible link has a fuse in the cell structure that is blown during programming when a higher than normal working current (I) is switched through the transistor and hence

84

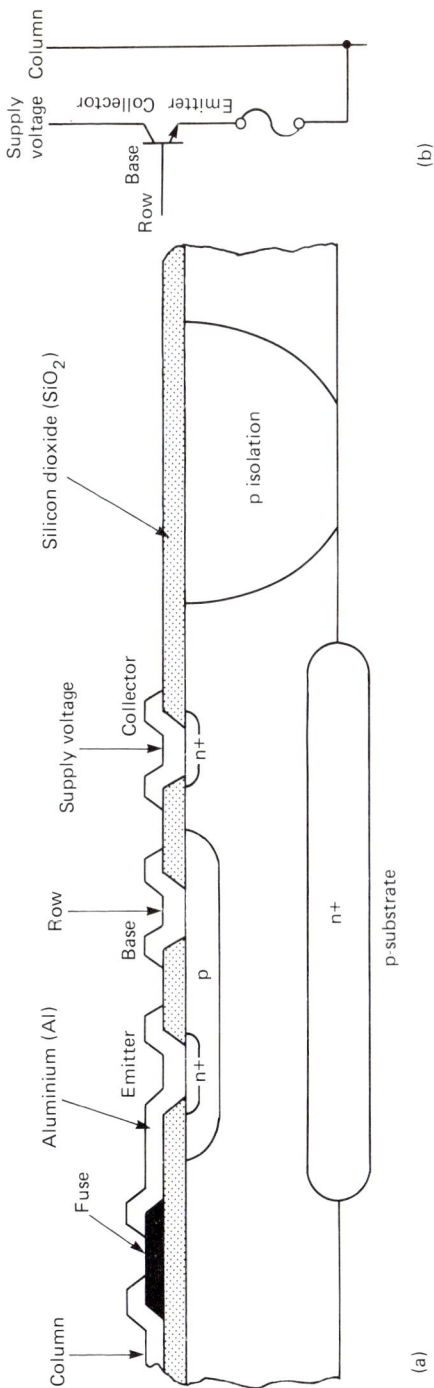

*Figure F.5* (a) A cross-sectional view of the basic structure of a bipolar fusible link (FL) memory, (b) logic elements

85

through the fuse in the emitter leg. Fusible link memory devices are completely non-volatile and once programmed cannot be erased. *See Figure F.5.* Continued improvements in fuse technology have resulted in four kinds of currently employed fuse materials, polycrystalline silicon, platinum silicide, nichrome and titanium tungsten.

**FLOP** FLOating Point. A notation to represent decimal numbers that is employed to help simplify arithmetic calculations by moving the decimal or binary point from left to right. 568 000 000 can be shown as 5.68, 8 as it is equal to $5.68 \times 10^8$.

**FLOTOX** FLOating gate Tunnel OXide. A principle employed in electrically-erasable programmable read-only memories

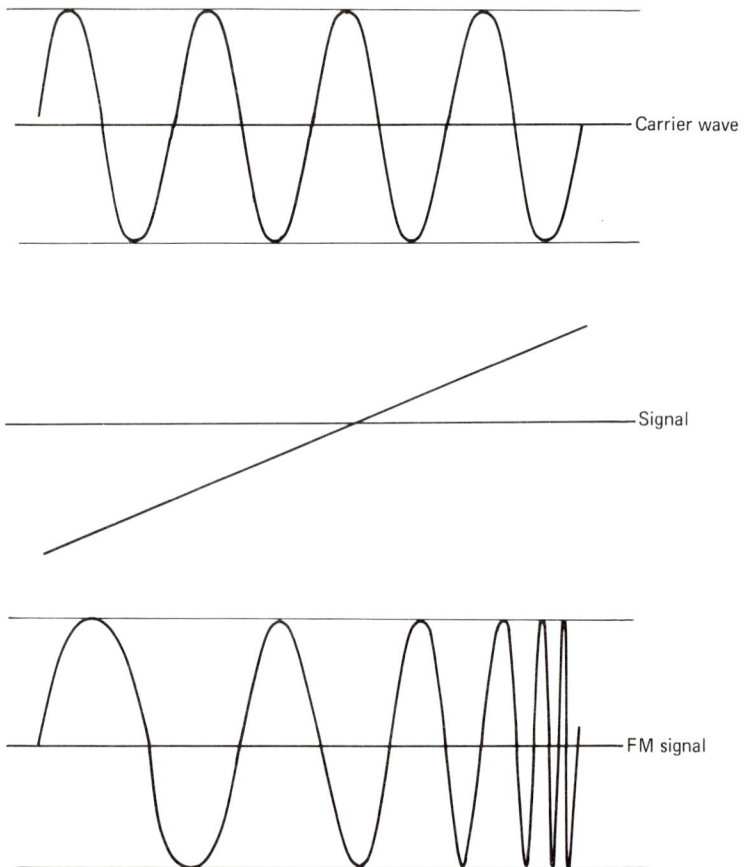

Carrier wave

Signal

FM signal

*Figure F.6*  Frequency modulation (FM)

(EEPROM) in which electrons tunnel to and from a floating gate. *See Figure E.2*, p. 71.

**FM** Frequency Modulation. A method of modifying a sine wave signal so that it carries information, as shown in *Figure F.6*. The sine wave or carrier wave has its frequency modified in accordance with other information required to be transmitted. The frequency function of the modulated wave may be continuous or discontinuous. In the discontinuous case two or more particular frequencies may each correspond to one significant condition.

**FMS** (1) File Management System. A collection of computer programs or an on-line executive program that provide the clear management of files. They allow for the creation, deletion or retrieval of files by name from a bulk storage device.

(2) Flexible Manufacturing System. Controlled by numerical control (NC) machines linked into a computer numerical control (CNC) system, an FMS makes use of robot-controlled transport of production work between machine stations.

**FO** Fibre Optics. The activity and technology of the passive guidance of ultra-violet (UV), visible or infra-red light through predetermined paths along transparent flexible fibres. These fibres provide a path for a single beam of light or in multiples, such as the transportation of a complete image. The fibres are provided either singly or in a cable bundle.

**FOC** Fibre-Optic Communications. A communications system that provides for the carrying of a signal from one point to another by the use of optical fibres (OF) or fibre optics (FO).

**FORTRAN** FORmula TRANslator. A computer programming language originally conceived by International Business Machines (IBM) and developed for use on scientific analysis and mathematical problems. It is now extensively used for commercial and business applications. It is a widely used language in scientific environments.

**FOSDIC** Film Optical Sensing Device for Input to Computers. A device to enable a computer memory to receive information from a microfilm storage medium.

**FOTS** Fibre-Optic Transmission System. A system of transmission that uses small-diameter transparent fibres through which ultra-violet, visible or infra-red light is passed. *See* FO, fibre optics.

**FPLA** Field Programmable Logic Array. An arrangement of logic circuits in which the internal connections of the gates can be programmed by the use of a higher than normal working current (I), and can therefore be programmed by the user. Fusable links (FL) are used for programming the logic configuration and the high current is passed through the links to achieve the required logic design.

**FPP** Floating Point Package. A set of computer software routines that are necessary to perform floating point (FLOP) arithmetic.

**FPS** Feet Per Second. A measure of speed, often used to indicate the rate at which a magnetic tape (MAGTAPE) passes a magnetic head.

**FR** Forced Release. A facility in a telecommunications system that enables a traffic carrying device or circuit to be released arbitrarily, quite independently of any release action taken by the caller or the called party.

**FROM** Field programmable Read-Only Memory. A read-only memory (ROM) that is programmed by fusing or burning out particular critical interconnections at specific storage locations. (*See* FL, fusible links). The programming can be completed by the customer using special equipment.

**FS** (1) File Separator. An indicator or flag employed to show the boundary between one collection of related records treated as a whole unit (file) and another.

(2) Final Selector. A telecommunications switching unit that has one inlet and provides direct access to a number of subscribers' lines via its outlets. Operation can be by external or internal controls.

(3) Field Separator. A control character employed to indicate or flag the boundaries of a group of contiguous locations (or columns on a card) which contain a particular item of data or a group of related pieces of information.

**FSD** Full-Scale Deflection. (a) A term normally applied to an analogue meter, in which the indicating pointer of the device reaches the maximum value for which the instrument has been scaled. (b) The absolute maximum value or full scale value of a measured amount of whatever specific quantity for which a particular measuring instrument has been calibrated.

**FSK** Frequency-Shift Keying. A form of frequency modulation (FM) where the frequency is made to vary without interrupting the carrier at the instant limiting intervals of modulation (i.e. the significant conditions of the modulating digital signal), between a set of predetermined discrete values called the 'mark' and the 'space' frequencies. Generally FSK is used in telecommunications for lower speed devices such as teleprinters.

**FSL** Formal Semantics Language. A language used for computer-orientated languages that is employed as a machine-independent language. It intergenerates compilers for a specific machine and contains formal semantics.

**FSP** Frequency-Shift Pulsing. A technique of signalling, used in telecommunications, which makes use of a frequency shift between 1070 Hz and 1270 Hz. This particular type of signalling is used with narrow-band systems, such as teletypewriter (TTY) switching networks.

**FSU** Final Signal Unit. A group of information and check binary digits (BITs), normally subject to a single error detecting process, that are used in the common-channel signalling (CCS) or multi-unit messages. The International Telegraph and Telephone Consultative Committee (CCITT) have designated a signal unit system No. 6 which consists of 28 bits, the last being used for error-detection purposes.

**FTET** Full Time Equivalent Terminals. The access to a computer system of a particular terminal can be quantisized by using FTET. It allows for the number of hours that the terminal is available for direct on-line access.

**FTP** File Transfer Protocol. Permitting transfer between different computers, FTP is a formalization of the methods by which the two parties communicate.

**FTS** Federal Telecommunications System. A leased telephone network in the USA shared by federal government agencies.

**FX** (1) Foreign eXchange. Provides connection between a customer and a remote telephone exchange, giving the equivalent of local service from the distant exchange.
(2) FiXed area. A portion of a computer disk where core image programs can be stored and protected.

# G

**g** Gram. The unit of mass in the centimetre, gram, second (CGS) system of units, it is defined as one thousandth of a kilogram (Kg).

**G** (1) Giga. A decimal prefix commonly used in association with a base unit in the SI system of units, indicating a multiplication factor of $10^9$ of that unit. For example, one gigahertz (GHz) is equal to 1 000 000 000 hertz.
(2) Gauss. In the centimetre, gram, second (CGS) system of units the gauss is the unit of magnetic flux density. Now superseded by the tesla (T), one gauss equals $10^{-4}$ tesla.
(3) Generate. An output (O,2) from an adder in a microprocessor (MP) that has been designed for the connection to a carry look ahead circuit.
(4) Ground. Also known as 'earth'. An intentional or accidental

connection made to earth, or a large conducting body. The earth connection provides a reference potential level and is also employed to carry earth currents in communication circuits.

**G³** Gadolinium Gallium Garnet. A nonmagnetic substance, grown as single crystals. When cut into wafers, it is used as a substrate for solid-state magnetic circuits.

**GaAs** GAllium ArSenide. A semiconductor material that can be used for high-speed applications. It is difficult to work with and its main uses have been for microwave devices, but lately it has been employed for metal semiconductor field-effect transistors (MESFETs) used in high-speed logic circuits.

**GALPAT** GALloping PATtern memory. A microprocessor (MP) test technique for generating successive patterns.

**GAP** General Assembly Program. Allows programs to be written in symbolic code rather than in the absolute machine language of a computer.

**GB** Grid Base. The span or range of grid voltage (V) between zero and the voltage that provides a cut off of anode current in a gas-filled electron tube. It thus provides control of the current (I) flow between the two other electrodes.

**Gbps** GigaBits Per Second. An information transfer or signalling rate of $10^9$ binary digits (BITs) per second.

**GDF** Group Distribution Frame. In a frequency division multiplexing (FDM) system, GDF is a frame that provides for the interconnection apparatus used in the twelve transmission channels (CHNL(a)) assembled in frequency band 60–108 kHz.

**GDO** Grid DipOscillator. Used for the measurement of radio frequency (RF), this portable instrument employs a grid-current meter and an oscillator. Resonance in an antenna is detected when the meter indicates a sharp dip.

**GDT** Graphic Display Terminal. A visual display unit (VDU) that permits the user or operator to display graphics information.

**GE** (1) Greater than/Equal to. A comparison of two values that states whether one value is greater than or equal to another. Normally it is used in order to allow another process or calculation to be carried out.

The symbol for GE is $\geqslant$.

(2) Gateway Exchange. A telephone exchange that allows the connection of international exchanges with a national network.

**GEORGE** GEneral ORGanizational Environment. An organized collection of techniques and procedures for operating a computer, that resides on a disk, employed with ICL computers.

**GHz** GigaHertZ. A frequency of one thousand million hertz (Hz).

**GI**   Graded Index. Characteristic of an optical fibre (OF) which has a refractive index that varies and is lower at distances radially away from its axis. This causes light rays to be continually refocussed by refraction in the core. *See Figure G.1.* It bends the rays inwards and allows them to travel faster in the lower index of refraction region. A GI fibre provides high band-width capabilities.

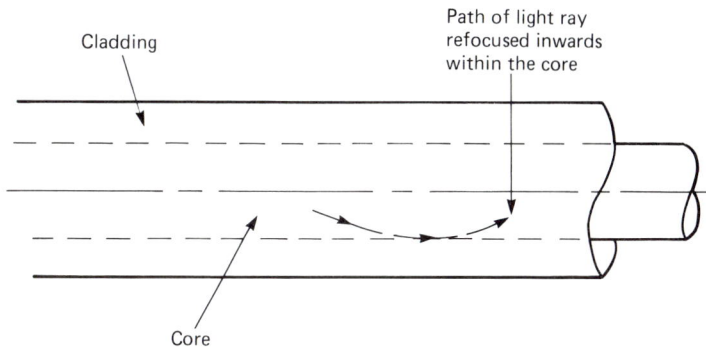

*Figure G.1*   Graded-index (GI) optical fibre (OF)

**GIGO**   Garbage In, Garbage Out. Produced as a result of undesirable input information or data, GIGO is unwanted and meaningless information carried along in the storage of a computer.

**GIRL**   Generalized Information-Retrieval Language. A computer language employed in the searching of stores of data held on computer files. It has been developed by the US Defense Nuclear Agency.

**GJP**   Graphic Job Processor. A feature of a computer program for job control, that allows job control statements and other associated information to be entered via a remote visual display unit (VDU).

**GKS**   Graphical Kernel System. A standard defined in ISO GKS 7.2 for computer systems. Comprising a set of basic functions for graphical input/output (I/O) on vector and raster devices, GKS makes for portability of application programs between installations, savings in programming effort and simplified viewing of a variety of display devices. It can support several work stations in parallel.

**GM**   Group Mark. A special character used in a write instruction to designate the beginning or end of a group of words, characters or records in storage. It enables the group so identified to be processed and addressed as a unit.

**GND**  GrouND. Usually connected to earth for safety, the GND is a point on a circuit at zero volts (V) potential. All points on a circuit at this zero potential can be said to be connected by a common ground.

**GOS**  Grade Of Service. A measure of service provided by a telephone exchange during a period of an hour in which the average traffic intensity is at maximum. It is usually expressed as the ratio of telephone calls liable to failure at the first attempt due to equipment limitations.

**GPC**  General-Purpose Computer. A generic term for computers (CMPTs) that can be used for a number of purposes, rather than dedicated to a specific function or functions.

**GPIA**  General-Purpose Interface Adaptor. Able to interface between an IEEE-488 standard instrument bus and a microprocessor unit (MPU), the GPIA enables many instruments to be interconnected and thus remotely automatically controlled or programmed.

**GPIB**  General-Purpose Interface Bus. A bus that connects instruments, including the connectors, the signal lines and the voltage level on lines, the functions used by the instruments interfaces to exchange data and control messages and protocols. The mechanical specification defines the cable and connectors and the electrical specification defines the voltage and current values required to be compatible with transistor transistor logic (TTL).

**GPO**  General Post Office. UK organization now split into two parts – the Post Office (PO), which handles postal service, and British Telecom (BT), which handles telephones and telecommunications.

**GPSDW**  General-Purpose Scientific Document Writer. Capable of producing complex hard copies for scientific use, this computer output (O,2) device can deal with mathematics and graphs.

**GPSS**  General-PurpoSe Simulation program. A high level computer programming language. Each natural language instruction corresponds to a number of machine code instruction. It permits the use of notations with which the user is already familiar.

**GRP**  Group Reference Pilot. A number of frequency waves transmitted to help maintain and adjust a system. It is applied where telephone channels (CHNL(a)) are assembled into groups, accompanying the group until they are separated into their individual channels.

**Gs** Gauss. In the centimetre, gram, second (CGS) system of units the gauss is the unit of magnetic flux density. Now superseded by the tesla (T), one gauss equals $10^{-4}$ tesla.

**GS** Group Separator. A special character used to designate the beginning or end of a group of words, characters or records in a store. It enables the group to be identified and be processed and addressed as a unit.

**GSC** Group Switching Centre. A term used in the UK for a telephone exchange that connects a number of local exchanges to traffic routes, between exchanges more than 24 km (15 miles) apart.

**GSI** Graphic Structure Input. Permitting the direct searching of a store of information held on a computer file, GSI methods are used in connection with scientific material and data.

**GT** Greater Than. A comparison of two values that states whether one value is greater than another. Normally it is used to allow another process or calculation to be carried out. The symbol for GT is >.

**GTH** Gas-Tight High-pressure. A method of stacking printed circuit boards (PCBs) and providing efficient connection without the need for goldplating. The spring contacts dig relatively deeply into the softer PCB pad, to provide a gas-tight seal around the metal contact. The GTH principal has also been incorporated into the design of a printed circuit board (PCB)

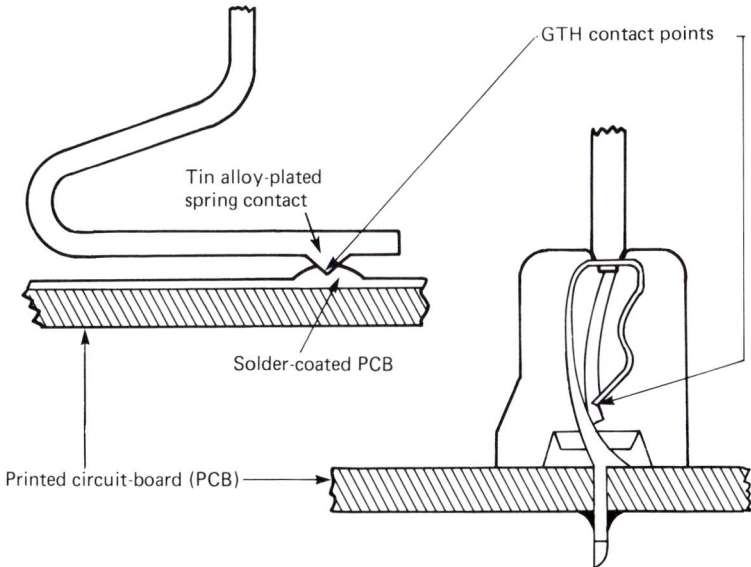

*Figure G.2*  Two examples of gas-tight high-pressure (GTH) contacts

mounted, zero insertion force (ZIF) termination for flat or desctete solid standed conductors. A simple example of each is shown in *Figure G.2*.

**GTN** Government Telecommunications Network. A network that links more than 400 governmment offices throughout Britain by means of tandem exchange and private circuits rented from British Telecom (BT).

**GUTS** Gothenburg University Terminal System. Designed and implemented by the University of Gothenburg, GUTS is a particular type of interactive system available on International Business Machines' (IBM) equipment and machines.

# H

**h** Hecto. A decimal prefix used in association with a base unit in the SI system of units indicating a multiplication factor of $10^2$ of that unit. For example, one hectometre hm is equal to 100 metres (m,2).

**H** (1) Henry. The SI unit of inductance and permeance.

(2) High. The most significant half (that containing the most significant digits MSD) of a microcomputer memory location or pointer. Normally, the 0–7 portion of a 16 binary digit (BIT) word.

(3) Quadripole. A prefix used in relation to quadripole network. It has a pair of inputs and a pair of output (O,2) terminals. *See Figure H.1.*

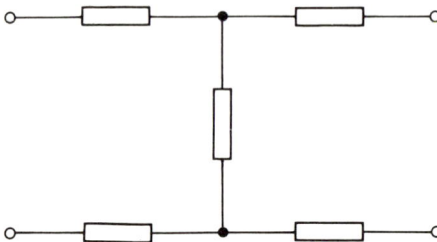

*Figure H.1*　A quadripole (H) network

**HA** Half Add. A computer instruction that carries out binary digit (BIT)-by-binary digit half additions. The operation is by a half-adder device where two inputs are received and two outputs (O,2) are produced representing the sum and the carry.

**HAL** Harwell Automated Library. A computerized system run by the Atomic Energy Authority at Harwell that provides for the loan of items of scientific data from the Authority's library.

**HAM**  A slang expression used to describe an amateur radio operator.

**HAMT**  Human-Aided Machine Translation. A computerized system capable of automatically translating text from one spoken language to another. It requires human intervention only to resolve ambiguities of syntax and semantics as well as problems arising from non-literal usage.

**HASP**  Houston Automatic Spooling Priority system. Providing spooling and priority functions for a computer operating system (OS) made by International Business Machines (IBM), this system has a number of modified versions.

**HC**  (1) Heat Capacity. A measure of the amount of heat that can be absorbed by a component. In the case of a diode (D,5) during operation, it would be the heat absorbed during pulses so short that little heat flows out of the diode into its heat sink during the pulse.

(2) Heat Coil. A thermally-operated protection device connected between a transmission line and a telephone exchange. It operates at temperatures that are too low for other protection devices such as fuses.

**HD**  Half-Duplex. (a) Generally in computer terminology, a circuit or system that can transmit in either direction, but not both directions at once. (b) A circuit designed for duplex (DX) operation, but limited by the nature of the terminal equipment to use in alternate directions only.

**HDB**  High Density Bipolar. A version of coding used in digital transmission. Long strings of zeros are replaced by coded groups containing timing information.

**HDLC**  High-level Data Link Control. A formalized method of communication in which information is considered as a stream of binary digits (BITs), allowing data transparency via bit stuffing.

**HDX**  Half-Duplex. *See* HD.

**HELP**  The HELP mode. A particular method of operation of a computer, terminal or processor. The user is guided by the HELP mode which indicates instructions on a visual display unit (VDU). The mode enables the user, with no programming experience, to perform functions or obtain information by using single keystrokes.

**HEX**  HEXadecimal. A system of number notation to the base 16. The most commonly used set of digits representing the hexadecimal notation system are 0 to 9 and then A, B, C, D, E and F. There are some other systems not so widely used that employ letters other than A-F for the last six digits (DGTs).

**HF**  High-Frequency. In radio systems, the HF bands lie between 1·6 and 30 MHz, and are popular because relatively long distances can be covered at low cost, under relatively light traffic conditions.

**Hg**  Mercury. A metallic element, atomic number 80, atomic weight 200.59, specific gravity 13.6, melting point −39°C and boiling point 357°C. It is used in thyratrons, arc rectifiers, tilt switches and to 'wet' the contacts of various switches.

**HICLASS**  HIerarchical CLASSification. One in which terms are provided with a designation and arranged in a hierarchical order. Items are split into initial sets, and then into successively finer sets. The Dewey decimal classification (DDS) system uses this method.

**HiNIL**  HIgh-Noise-Immunity Logic. A specific design of logic circuits for use in applications where transients are likely to be present producing a noisy environment. They can protect complementary metal-oxide semiconductor (CMOS) circuits from transients and static electricity, and are compatible with a number of analogue circuits.

**HISAM**  Hierarchical Indexed Sequential Access Method. A technique of organizing a store of information entered on a computer file, that can be held on a disk.

**HJ**  Hetero Junction. Refers to devices that have several deposited layers and hence a number of p-n junctions (p-n).

**HLL**  High-Level Language. A computer programming language resembling natural language using symbols and command statements that an operator can read, but is not English. Each statement represents a powerful series of computer instructions. Examples of high-level languages are: Beginners all-purpose symbolic instruction code (BASIC), common business-orientated language (COBOL) and formula translator (FORTRAN).

**HMOS**  High-speed Metal-Oxide Semiconductor. A metal-oxide semiconductor (MOS) technology that employs high-scaling techniques to increase density (D,3) and performance beyond that of n-channel semiconductor (NMOS) and p-channel metal-oxide semiconductor (PMOS) technologies.

**HNIL**  High-Noise Immunity Logic. See HiNIL.

**HOL**  High-Order Language. The same as high-level language (*see* HLL).

**HP-IB**  Hewlett-Packard Interface Bus. An implementation of IEEE-488 (*See* IEEE-488) standard for digital interfacing of programmable instrumentation made by Hewlett-Packard.

**HRC**  High Rupturing Capacity. Usually refers to a type of cartridge fuse which has a high performance. Its minimum

rupturing capacity is above 16 500 amps (A) and it is designed to
cut off rapidly.

**HSM**  High-Speed Memory. A computer peripheral device that is
capable of retrieving information at a comparatively faster rate
than other memory units connected to a computer system.

**HSP**  High-Speed Printer. A printer that is able to operate on line
and has a speed of operation compatible with data processing
(DP) and computer speeds.

**HSR**  High-Speed Reader. A device that converts information
from one storage medium into another. In general a card reader
operating at speed more compatible with the speed of
computation and data processing (DP) so that it may operate
on-line.

**HSYNC**  Horizontal SYNChronization. A signal employed in a
television (TV) set to determine the horizontal position of the
picture image on the screen.

**HT**  (1) High Tension. A high-voltage supply required for anodes
and screen grids of electron tubes. Usually in the range of
60–250 volts (V).

(2) Horizontal Tabulation. A specific control character that
causes the position of print of a printer to be pushed forward to
the next allocated position in the same line of print.

**HTL**  High Threshold Logic. A type of microcomputer logic
employed in industrial environments, where the electrical noise

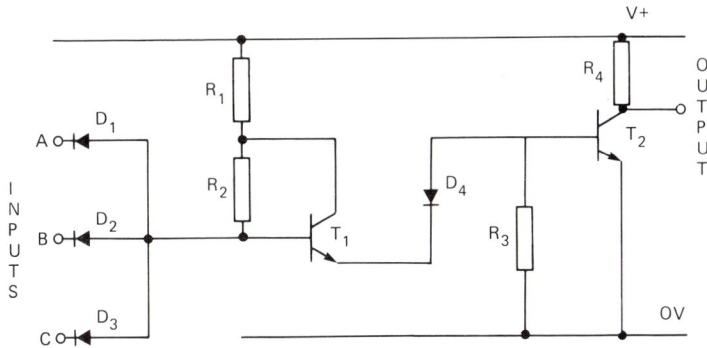

*Figure H.2*  A typical high-threshold logic (HTL) gate

level can be quite high. Digital systems operated in such
environments must have a high threshold in order to ignore
these noise spikes, and thus not provide false operation. The
HTL gate is a development of diode transistor logic (DTL). A
typical HTL gate is shown in *Figure H.2*.

**HUD**  Head-Up Display. An arrangement of equipment that projects images into a pilot's line of sight. It uses a cathode-ray tube (CRT) to provide comprehensive information on height, speed, heading and altitude.

**HV**  High Voltage. A term used in reports and documents which generally refers to a voltage of more than 650 volts (V).

**HVDC**  High-Voltage Direct Current. A direct current (DC) with a voltage in excess of 650 volts (V).

**HVPS**  High-Voltage Power Supply. A form of power-supply unit (PSU) that can provide a high-voltage (HV) supply output (O,2) from a lower-voltage input. A typical HVPS is found within a domestic television (TV) receiver and is used to provide the cathode-ray tube (CRT) element voltages.

**HYPERNET**  HYPER NETwork. A generic term used to describe a number of computers interconnected to form a network capable of transmitting information at megabaud rates.

**Hz**  HertZ. The SI unit of frequency equal to the number of cycles per second (C/S). For example, a frequency that has 3 complete cycles in a second is said to be at 3 hertz.

# I

**I**  Current. The rate of flow of electricity past a given point. The unit of measurement is the ampere (A).

**IACS**  International Annealed Copper Standard. A standard of purity and conductivity laid down by the International Electrotechnical Commission (IEC). Annealed copper is said to have a conductivity of 100% IACS.

**IAD**  Initiation Area Discriminator. A type of specialized cathode-ray tube (CRT).

**IAL**  International Algebraic Language. A computer language which was the predecessor of algebraic-orientated language (ALGOL).

**IAM**  Initial Address Message. The first signal message transmitted in a telecommunications call set-up process, in common-channel signalling (CCS). It contains all the information necessary to identify the called party or the route to be taken.

**IAR**  Instruction Address Register. A device that temporarily stores the location in a memory (address) of the next instruction to be executed.

**IBA**  Independent Broadcasting Authority. So called because of its independence from any control or finance provided from licence fees (as compared to the BBC). The IBA controls all UK commercial television and commercial sound radio.

**IBIS** International Book Information Service. Providing information for Prestel, this service's major activity is sending information concerning recent publications to subscribers. The subscribers' particular interests are held on a computer file.

**IBM** International Business Machines. The largest manufacturer of computers in the world.

**IC** (1) Integrated Circuit. (a) A single-package complete circuit (CCT), able to perform at least one, sometimes many, complete functions. This interconnected array of components is fabricated from a single crystal of semiconductor material by etching, doping and diffusion. Generally, ICs are known by their circuits' complexity as follows: small-scale integration (SSI), medium-scale integration (MSI), large-scale integration (LSI), very-large-scale integration (VLSI) and extra-large-scale integration (ELSI). (b) IC represents integrated circuits on parts lists and schematic diagrams.

(2) Instruction Counter. A multibit device which temporarily stores information that monitors the location in a memory (address) of the current instruction and is used to input information in the memory-address register (MAR).

**ICA** Integrated Communications Adaptor. An integrated adaptor that permits connection between multiple communication lines and a processing unit.

**ICAND** MultiplICAND. One quantity or amount that is multiplied by another quantity or amount.

**ICC** International Computation Centre. The computer centre's services are available to member nations of its sponsoring organization UNESCO.

**ICE** (1) In Circuit Emulator. A device that actually replaces a microprocessor (MP) by connecting directly to the circuit board (in circuit). The device has a hardware/software capability for the real time input/output (I/O) debugging of chips (CHIPs). The emulated microprocessor can be stopped, its memory locations can be examined and modifications as well as other alterations can be carried out.

(2) Interface Cancellation Equipment. An electronic counter-counter measure (ECCM) system for all types of very-high-frequency (VHF) combat radio. The system can upgrade current equipment to meet electronic counter measures (ECMs) used against battlefield VHF communications.

**ICIREPAT** International Co-operation Information Retrieval among Examining PATent offices. This Geneva-based organization promotes international co-operation between numerous national patent offices and is concerned with the documentation and retrieval systems run by the various offices.

**ICL** International Computers Limited. The largest manufacturer of computers in the UK.

**ICS** Integrated Communication System. A system incorporating two or more autonomous communication systems combined into one single system, so that none of the original autonomy remains.

**ICT** In Circuit Test. The examination of discrete components and integrated circuits (IC) as part of a complete circuit (CCT). This can be done statistically in a non-energized state and dynamically in an energized state. The term is used generally in association with tests via automatic test equipment (ATE).

**ICU** Instruction Control Unit. Typically this device contains a read-only memory (ROM) storing the microinstructions and address control logic for branching. This part of a computer enables instructions to be retrieved in the correct order, interpreted and applied.

**ICW** Interrupted Continuous Wave. A type of controlled disturbance propagated through space (wave), used for radio telegraphy where the audio-frequency (AF) wave is keyed on and off.

**ID** (1) Insulation Displacement. The technology employed in insulation displacement connectors. (*See* IDC.)

(2) IDentification. A computer, data processor (DP) or telecommunications label that can consist of a code number or code name which serves to uniquely identify a record, block, file or other unit of information or data.

**IDC** Insulation Displacement Connector. A type of connector and connector system that provides a self-stripping contact. The contacts are normally in the shape of a 'U' formed by two parallel cantilevered beams or legs with sharpened tips. The majority of flat cables for use with IDC systems are manufactured on 0.050-inch centres. A complete flat cable with as many as 64 conductors can be terminated in one process.

**IDDD** International Direct Distance Dialling. An automatic system of calling telephones beyond the local area, enabling subscribers to make international calls. *See* DDD.

**IDF** Intermediate Distribution Frame. A unit that provides for the interconnection, termination or cross-connection of cables and wires from units of telephone exchange equipment.

**IDN** Integrated Digital Network. A network using both digital transmission and digital switching.

**IDP** (1) Integrated Data Processing. (a) A computer system processing concept in which the processing of data is organized, correlated and systematic, embracing all the data processing (D,1) requirements in a particular area of interest. (b) Treating data processing requirements as a whole, this system tries to reduce or eliminate the duplication of data entry or processing steps. (c) A system where all procedures are linked by a computer employed for the processing of data.

(2) InterDigit Pause. An interval of time experienced when telephone dialling, between the end of one set of pulses which represent a digit (DGT) and the start of the next.

**IDPM** Institute of Data Processing Management. A professional body based in the UK, formed from the Institute of Data Processing and the Data Processing Management Association.

**IDU** IDle signal Unit. A group of information and check binary digits (BITs) used in common-channel signalling (CCS) during a period when no signal messages are free for transmission.

**IEC** International Electrotechnical Commission. Probably best known for the formulation of the MKS (metre, kilogram, second) system of fundamental units, the IEC, which is based in Geneva, has produced much standardization in the electrical field since its formation in 1906. It recommends standards for electrically-operated equipment, working with the International Standards Organisation (ISO) in this particular area.

**IECQA** International Electrotechnical Commission Qualification Assessment. Operating under the authority of the IEC, the IECQA defines and implements quality assessment procedures for electronics components, which are not mandatory for members of the IEC.

**IEE** Institution of Electrical Engineers. A UK professional body that promotes the general advancement of electrical/electronics science and engineering and their application: facilitates the exchange of information and ideas on these subjects; and financially aids the promotion of invention and research. It is also concerned with the standardization of electrical/electronics equipment and offers to the British Standards Institution (BSI) advice for determining national standards.

**IEEE** Institute of Electrical and Electronics Engineers. (US). A professional body concerned with the standardization of electronic equipment. It also determines standards on behalf of the American National Standards Institute (ANSI).

**IEEE-488** A standard of the IEEE that is widely used to specify a microprocessor (MP) parallel bus ideal for high-speed module-to-module communication as well as for system-to-system communication. *See* GPIB.

**IEEE-583** Usually called CAMAC, this is a standard laid down by the Institute of Electrical and Electronics Engineers (IEEE). It is a hardware/software specification that was originally developed for the nuclear industry but is currently widely adopted by many other industries. *See* CAMAC.

**IEEIE** Institution of Electrical and Electronics Incorporated Engineers. A UK professional body that provides for electrical and electronics engineers and technicians a recognized engineering qualification of a high level.

**IERE** Institution of Electronic and Radio Engineers. A UK professional body that promotes the general advancement of the science and practice of radio and electronics engineering, and facilitates the exchange of ideas and information on these and other branches of engineering and their applications.

**IEV** International Electrotechnical Vocabulary. A standardized set of electrical terms, devised by the International Electrotechnical Commission (IEC).

**IF** (1) Intermediate Frequency. A carrier frequency to which the modulation of all received signals is transferred in a superheterodyne radio receiver. It appears at the output (O,2) of the frequency changer. The transfer is achieved by a frequency changer circuit (CCT).

(2) A statement used in many computer languages. A reserve word signifying the beginning of a conditional statement.

**IFCS** International Federation of Computer Sciences. A collection of organizations concerned with the computer sciences, one representing each member country.

**IFF** Identification Friend or Foe. A system used to indicate whether an aircraft is an ally or an enemy. In operation the friendly aircraft is equipped with a transponder that replies with a coded signal to an interrogation initiated by the IFF system.

**IFIPS** International Federation of Information Processing Societies. A collection of organizations concerned with information processing, one from each member country.

**IFRB** International Frequency Registration Board. A permanent body of the International Telecommunication Union (ITU) established in 1947. By maintaining a master list of radio frequencies (RF), it tries to prevent a country using a new frequency that would interfere with existing radio services.

**IF-THEN-ELSE** A program statement often used in high-level computer languages. If the IF statement is correct, a THEN statement has the meaning that a following expression is to be executed. When the expression is untrue, the ELSE expression is executed.

**IGFET** Insulated-Gate Field-Effect Transistor. A field-effect transistor that is constructed with a thin dielectric barrier, used to isolate the input electrode (gate) from the channel (CHNL(h)). *See Figure 1.1.* When a control voltage is applied to the gate, it induces an electric field across the dielectric barrier and modulates the nature of the channel region. There are two basic structures of IGFETs, p- and n-channel devices, and there are also two basic classifications of modes of operation, *enhancement* and *depletion*.

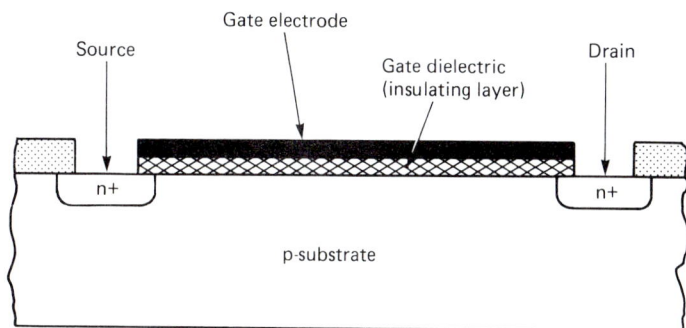

*Figure I.1* A cross-sectional view of an insulated-gate field-effect transistor (IGFET)

**IIC** International Institute of Communications. Providing analysis of issues relating to communication and particularly electronic communication, the IIC is UK-based but has international connections. It aims to provide governments and industries with relevant information.

**IIL** or **I²L** Integrated Injection Logic. A family of bipolar integrated circuits (BIC) evolved from diode transistor logic (DTL). They are very compact and provide high-speed operation with a high functional packing density. *See Figure I.2.* I²Ls use less power, are more stable and cheaper, can be easily mass-produced and are two orders of magnitude faster than transistor-transistor logic (TTL).

**IIR** or **I²R** Rate of energy loss. In an electrical conductor of resistance R carrying a current, I, I²R is an expression of the rate of energy loss, i.e. the heating effect of a current. It also represents the energy loss.

**IL** IdLe. A period during which a computer (CMPT) or system stays switched on and available but is not in operation.

**I²L** *See* IIL.

**ILD** Injection Laser Diode. A semiconductor diode with a more directional light emission pattern and a narrower spectral width than a light-emitting diode (LED). It is ideally suited to, and commonly used for, optical communications.

**ILL** Inter-Library Loan. Refers to the information interchange between libraries. The information can be in the form of computer hard copy or computer output microfilm (COM).

**ILLIAC** ILLinois Institute of Advanced Computation. A faculty of the Univerisity of Illinois concerned with the design and implementation of a number of advanced and current technology computers.

**ILS** Instrument Landing System. Aircraft landing system employing line of sight (LOS,2) frequencies. It incorporates a localizer providing left/right guidance, a guide slope providing up/down guidance and a marker beacon to define progress along the approach course.

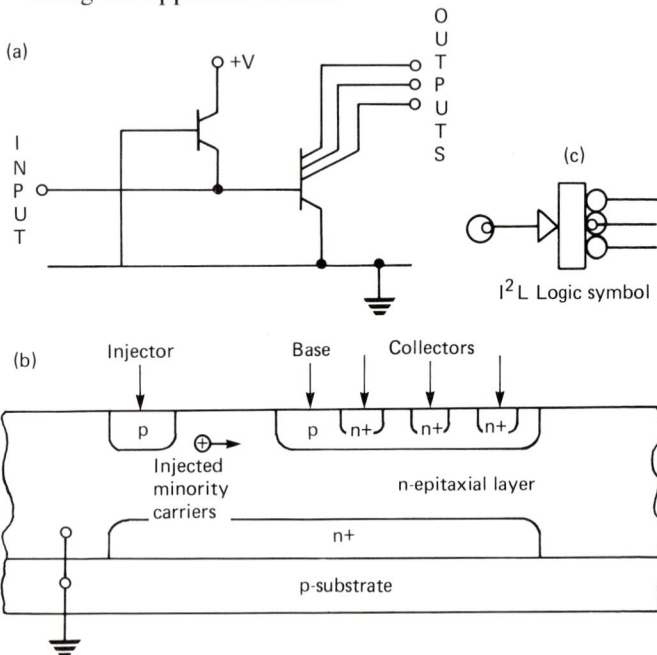

*Figure 1.2* (a) Equivalent circuit of a single integrated injection logic ($I^2L$) gate. (b) A cross-sectional view of a basic integrated injection logic-gate structure. (c) $I^2L$ logic symbol

**IM** (1) Item Mark. A special character or item of information used as an indicator to designate the beginning or end of a set of one or more characters (field) that have related data or information. The information concerning a particular object, event, transaction or operation.

(2) Intensity Modulation. In an IM fibre-optic communications (FOC) system, information is transmitted when a modulating electrical signal provides optical sources of variable intensity. De-modulation is achieved with a photo-detector whose electrical output (O,2) is proportional to that of the optical input.

104

**IMP** Interface Message Processor. A computer that transmits data by means of addressed packets that occupy the channel (CHNL(b)) just for the time taken to transmit the packet. It is used in an experimental data communication network funded by the Advanced Research Projects Agency (ARPA).

**IMPATT** IMPact ionization Avalanche Transit Time. A diode that acts as a powerful source of microwave ($\mu$W) power. The diode (D,5) is mounted in a microwave cavity and, as part of an appropriate circuit, spontaneous oscillation will occur.

**IMS** Information Management System. A group of procedures organized as a system that is able to carry out a number of processes upon information, including maintenance and organization cataloguing, locating, storage and retrieval.

**INFCO** INFormation COmmittee of the International Standards Organisation. One of the committees that advise the International Standards Organisation (ISO).

**INFOCOM** INFOrmation COMmunications. A message switching service to be used intra-company, supplied in the USA by Western Union.

**INFOL** INFormation Orientated Language. A high level computer language (HLL) structured around the provision of information. It permits notation to be used by operators with which they are already familiar.

**INFRAL** INFormation Retrieval Automatic Language. A computer language that permits the user to produce lists of groups of documentation information from indexed data.

**INIS** International Nuclear Information Service. Offering on-line and off-line searching, INIS uses stores of information held on computer files that relate to aspects of nuclear science and technology.

**INSIS** INter-institutional integrated Services Information System. A proposed information system to combine a number of services, including electronic mail, facsimile (FAX), telephone, telex (TX), teleconferencing and word processing (WP).

**INSPEC** INformation Services: Physics, Electrical and electronics, and Computers and control. Available for on-line searching, these stores of information held on computer files contain data relating to physics, electrical, electronic, computer and control subjects.

**INT** INTerrupt. (a) Normally initiated by an external signal, a discontinuity in the normal flow of a system or routine, that can be resumed at a later time. (b) A specific control signal that diverts the attention of a computer from the main program.

**INTEL** INTegrated ELectronics. A major manufacturer of semiconductor devices. Formed in 1969 it produced the first microprocessor (MP).

**INTELSAT** INternational TELecommunications SATellite consortium. A group of governments and their particular telecommunications authorities formed in 1964 to create a worldwide satellite system. Now over a hundred countries are members. Its first satellite was INTELSAT I known as Early Bird.

**INTERMARC** INTERnational MAchine Readable Catalogue. A machine readable cataloguing (MARC) system that is made available to a number of countries.

**INWG** International Network Work Group. An advisory body within the International Federation of Information Processing Societies (IFIPS) that is concerned with research into network protocols and interworking.

**I/O** Input/Output. Relates to the devices, media or operations that are employed to pass information in or out, or both, of a computer or system.

**IOC** (1) Input/Output Controller. A device that controls interactions between a processing unit and input/output (I/O) equipment. There are two portions to the controller; one controls actions of tape, card and printer equipment and the other is designed for the random processing of records stored on direct access devices.

(2) Integrated Optical Circuit. Consisting of miniature solid state optical components, an IOC is a circuit (CCT) or group of interconnected circuits, generally manufactured on chips (CHIP).

**IOCS** Input/Output Control System. (a) Computer software specifically designed to control input/output (I/O) operations and communications between peripheral equipment and a central processor unit (CPU). The main areas of communications that are concerned include device allocation, error recovery and file identification. An IOCS comprises a set of computer routines that can also direct other functions such as error correcting checkpoint label processing and restart. (b) A number of routines which a system analyst or programmer can use in a source program for handling input and output for tapes, cards and printers.

**IOP** Input/Output Processor. A device that is capable of handling normal data input/output (I/O) control and sequencing.

**IOR** Input/Output Register. A memory location that accepts information from outside a computer at a particular speed and presents the information to a computer calculating unit at a normally higher speed.

**IP** Information Provider. In the context of viewdata systems, this is an organization or an individual that provides data for the systems store of information held on computer files.

**IPC** Industrial Process Control. A data acquisition and control system that can provide great flexibility, as it can accept various types of process data, and a number of output (O,2) signals and format of data that a computer may exercise.

**IPL** (1) Information Processing Language. A particular computer programming language that provides for the processing of data representing information and the defining of the meaning of the processed information.

(2) Initial Program Load (Loader or Loading). An initiation process that allows a computer program or operating system (OS) into a computer with data records. Once the self-loading routine is established in memory, the program can proceed under its own control, loading and executing other programs.

(3) Initialize Program Load. A routine of initialization used in various intelligent controllers and some computers.

**I-POT** Inductive POTentiometer. A precision-wound toroidal autotransformer having two sliding contacts, a main contact and an auxiliary contact. During use the auxiliary contact is set to an initial calibration, while the main contact produces an output (O,2) voltage (V) that is relative to a mechanical rotary input.

**IPS** Inches Per Second. A measure of speed, normally used to indicate the rate at which a magnetic tape (MAGTAPE) passes a magnetic head.

**IPSS** International Packet Switching Service. A public automatic switched data service that provides access between UK data terminals and systems overseas.

**I²R** *See* IIR.

**IR** (1) Infra-Red. Energy that radiates, mainly heat, at wavelengths greater than visible red.

(2) Instruction Register. A device that temporarily stores the current instructions which are being decoded and executed by the central processor unit (CPU) section of a computer.

(3) Information Retrieval. The methods, techniques and processes employed in searching particular computer files for specific items of information, in response to specific user needs.

**IRAS** Infra-Red Astronomical Satellite. A satellite in a 900-kilometre-high (560 miles) near-polar orbit, launched in 1983 and circling the Earth 14 times a day. It carries a telescope cooled with liquid helium to perform the first survey of the cosmos at infra-red (IR,1) wavelengths (from 8 to 120 microns).

**IRE** Institute of Radio Engineers. A professional body in the USA concerned with radio equipment.

**IRG** Inter-Record Gap. A gap of 0.5–0.75 inch between blocks of information on magnetic tapes (MAGTAPE).

**IRIA** Institut de Recherche d'Informatique d'Automatique. A French scientific research establishment.

**IRL** Information Retrieval Language. A computer programming language that is used for the cataloguing of large amounts of data all related to one specific area of interest. It allows for the retrieval of all or part of this data quickly and precisely.

**IRS** Information Retrieval System. A system which provides particular items of data requested by users. Modern systems employ files of information that can be searched for specific items, in response to the definitions of particular user needs. *See* CAR,2, computer-aided retrieval and DRS, document-retrieval system.

**IS** (1) Information Separator. A special character used to designate the logical boundaries between items of information in data records.
(2) Indexed Sequential. A method in which computer files are organized on a direct access (DA,2) storage device. A directory is compiled with the address of each record along with its identification, thus permitting the fast access of a relevant record.

**ISAM** Indexed Sequential Access Method. A method that provides keyed access for rapid and direct retrieval of information from a fast and continuously updated file of information in a computer system.

**ISAR** Information Storage And Retrieval. (a) The process of storing and retrieving data. *See* IR,3, information retrieval. (b) A computerized system that provides information to users in response to requests.

**ISB** Independent Side-Band transmission. The transmission of the frequency bands above and below a carrier frequency, each independent of the modulating signal of the other side band. The system is limited by the difficulty of separating the two signals at the receiver.

**ISBD** International Standard Bibliographic Description. A convention employed in describing documents that is recognized internationally.

**ISBN** International Standard Book Number. A number used to aid the location of a particular book within a library or information retrieval system. Every book published is allocated its own unique number, consisting of ten digits (DGTs).

**ISDN** Integrated Services Digital Network. (a) A planned UK telecommunications network intended to provide a wide, flexible range of facilities for voice, data and image translation.

It envisages communications being at a standard speed of 64 000 bits per second (BPS). *See* LCDDS, leased circuit digital data service. (b) An integrated digital network (IDN) used for multiple services such as telephone and data transmissions.

**ISL** InterSatellite Link. The transmission of messages between communications satellites.

**ISM** Industrial, Scientific and Medical. A prefix applied to apparatus used in connection with radio interference studies to describe industrial, scientific and medical apparatus that produce radio frequency (RF) energy.

**ISO** International Standards Organisation. A specialized agency of the United Nations, founded in 1947, and concerned with international standardization across a wide variety of industries and their products.

**ISR** Information Storage and Retrieval. (a) The process of storing and retrieving data. *See* IR,3, information retrieval. (b) A computerized system that provides information to users in response to requests. The processes employed in the system use the searching of computer files for particular items of information, in response to specific user needs.

**ISSN** International Standard Serial Number. A number used to aid communication between periodical publishers and libraries, to assist inter-library loan (ILL) and maintenance of computer files. Every title published is given a unique number consisting of eight digits (DGTs) in two groups of four.

**ISU** Initial Signal Unit. The first of a group of information and check binary digits (BITs) of a multi-unit signal message used in common-channel signalling (CCS).

**IT** Information Technology. The techniques associated with handling information. They encompass the use of telecommunications systems and computers in handling and processing data, including the application, acquisition, storage and dissemination of pictorial, audio and numerical data.

**ITC** International Teletraffic Congress. Formed mainly from representatives of telecommunications authorities and industries worldwide, this organization serves as an arena for those involved with teletraffic.

**ITF** Interactive Terminal Facility. The capability of a computer system to allow a user at a local or remote terminal to elicit a response to any enquiry made of the system. It also allows for a continuous dialogue between the user and the system and provides for immediate diagnostic assistance to any errors made by the user.

**ITS** (1) Invitation To Send. A computer process which establishes contact with a terminal to allow the terminal the opportunity to send or transmit a message or information if it has any prepared.

(2) Insertion Test Signal. Test signals introduced to unused lines during the time interval between two fields in which the television (TV) picture is suppressed (field blanking) to monitor, test and measure the working of the system and its networks.

**ITT** International Telephone and Telegraphs. A multi-national company involved with electronics and telecommunications.

**ITU** International Telecommunication Union. A United Nations telecommunications agency that has been established to provide standardized communications procedures and practices including the allocation of broadcasting frequencies and radio regulations on a worldwide basis.

**ITV** Instructional TeleVision. Usually distributed over specially allocated broadcast television channels (CHNL(g)) in the US, ITV can also be supplied by closed circuit television (CCTV).

**IVPO** Inside Vapour-Phase Oxidation. A chemical vapour-phase oxidation (CVPO) process that is used for the production of optical fibres (OF). Dopants are burnt with oxygen and gas to form an oxide stream that is deposited onto the inside of a rotating glass tube. It is then sintered to produce a doped layer of a higher refractive index and drawn into a solid fibre.

# J

**J** (1) A symbol used to represent the imaginary component of the sum of a real quantity or number and an imaginary quantity or number.

(2) A J-antenna is a radiating conductor with a length of about half a wavelength of the radiating frequency. It is fed at one end by a matched feeder a quarter of a wave in length.

(3) Joule. In the SI system of units, the unit of energy or work done. It is defined as the work done when a force of one newton (N,1) is deplaced one metre (m,2).

(4) Jump. A specific computer (CMPT) instruction that will cause the instruction at the memory location (address) specified in the jump, to be the next executed.

**JACK** A term applied to both a single-pin plug and its matching socket. The leads from the microphones and earphones of domestic tape recorders and record players are fitted at their other ends with jack plugs which can be inserted into the appropriate jack sockets on the equipment.

**JAN** Joint Army Navy programme. A programme set up in the USA to standardize on electronic and other military equipment.

110

**JCL**  Job Control Language. A computer language (designed by IBM) that interprets user instructions and processes them to instruct the central processor unit (CPU) what to do with the user's program.

**JDL**  Job Description Language. Similar to job control language (JCL), JDL is rather more general in that it covers all aspects of defining a task or job to a computer system.

**JEDEC**  Joint Electron Device Engineering Council. A body that helps determine standards on behalf of the USA Standard Institute (USASI).

**JFCB**  Job File Control Block. A set of memory locations containing information about a particular task or job. The information can be accessed by various programs, thus providing control of the operation of the programs.

**JFET**  Junction Field-Effect Transistor. A particular type of field-effect transistor (FET) where the gate forms a combination of p- and n-type semiconductor material (p-n) with the channel (CHNL(h)). A cross-sectional view of a JFET is shown in *Figure J.1*. The channel of the JFET can have more than one gate.

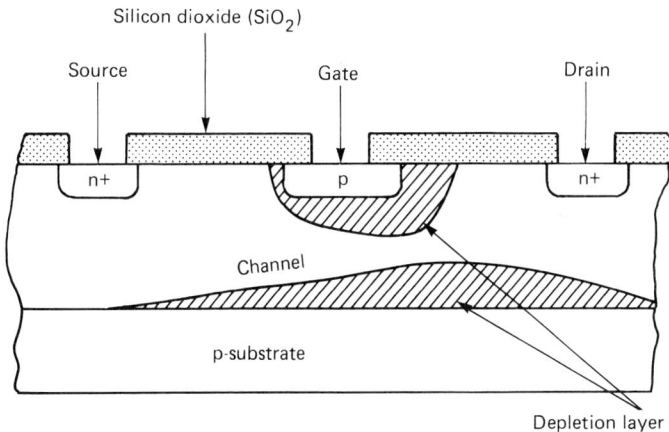

*Figure J.1*  Cross-sectional view of a junction field-effect transistor (JFET)

**JG**  Junction Grammar. A model of language employed in analysing sentences, it allows a sentence to be reduced into its syntactic elements. It is used in computerized and other techniques for the automatic translation of text from one spoken language to another; machine-aided translation (MAT).

**J-K**  A special form of flip-flop (FF) circuit that has two inputs, the J and K. *See Figure J.2.* At the application of a clock pulse

(CP), a 1 on the J input and a 0 on the K input will set the flip-flop at its 1 state. A 1 on the K input and a 0 on the J input will reset to the 0 state. A 1 simultaneously on both the J and K inputs will cause it to change state regardless of the previous state. A 0 on both the J and K inputs will prevent any change. Also known as a J-K bistable element.

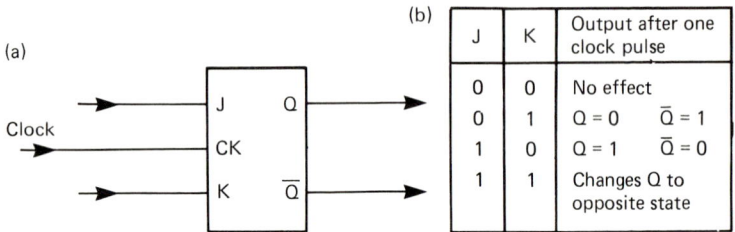

| (a) | | (b) | J | K | Output after one clock pulse |
|---|---|---|---|---|---|
| Clock / J Q / CK / K $\overline{Q}$ | | | 0 | 0 | No effect |
| | | | 0 | 1 | $Q = 0$    $\overline{Q} = 1$ |
| | | | 1 | 0 | $Q = 1$    $\overline{Q} = 0$ |
| | | | 1 | 1 | Changes Q to opposite state |

*Figure J.2*  The J-K flip-flop: (a) symbol, (b) truthtable

**JOHNNIAC**   JOHN von Neumann's Integrator and Automatic Computer. A computer based on the concepts of John von Neumann. Its particular feature is a program that is stored within the computer and executed under the control of a sequence register.

**JOSS**   Johnniac Open-Shop System. A time-sharing computer language modelled after formula translator (FORTRAN) or beginner's all-purpose symbolic instruction code (BASIC) that has been developed to make quick calculations that are far too complex for a calculator.

**JOVIAL**   Jule's Own Version of IAL. A particular algorithmic language (ALGOL) used by the US Air Force in some of their systems. *See* IAL.

**JUGFET**   JUnction-Gate Field-Effect Transistor. A field-effect transistor (FET) where the gate forms a combination of p- and n-type semiconductor material (p-n) with the channel (CHNL(h)). The channel of the JUGFET can have more than one gate. *See Figure J.1.*

# K

**k**   Kilo. A decimal prefix commonly used in association with a base unit in the SI system of units, indicating a multiplication factor of $10^3$ (1000) of that unit. For example, one kilometre (km) equals $10^3$ metres (1000 metres).

**K**   (1) A prefix used in computing to indicate a multiplication factor of $2^{10}$ (1024); e.g., one kilobyte (KB,3) equals $2^{10}$ bytes (1024).

(2) A non-standard but often-used symbol for kilohm; used especially on diagrams to replace the decimal point where this could be difficult to distinguish or cause confusion. 2.5 kilohms would be written 2K5, 2K5 ohms or 2K5Ω. The BSI standard representation would be 2.5KΩ.

(3) Kelvin. A temperature scale with a datum point set at absolute zero (where molecular motion ceases) which is 273.15 K below Celsius (C,8) freezing point, 0°. Used for temperature in the SI system of units. 1 K is an increment similar to 1°C as it has 100 units between the freezing and boiling points of water.

**KB** (1) Key Board. A unit consisting of a group of keys arranged in a particular order (*see* AZERTY, QWERTY, QWERTZ) and used to generate a set of signals for entering data or information. Alphanumeric keyboards may be used for word processing (WP), computer input, text processing and data processing (DP,1); numeric keyboards are used for calculators, touch-tone telephones and accounting machines.

(2) KiloBits. 1024 ($2^{10}$) binary digits (BITs).

(3) KiloBytes. 1024 ($2^{10}$) sequences of binary digits (BITs), normally consecutive 8 or 16 bits taken together, operated upon as a unit and usually shorter than a word.

**K band** K band microwave ($\mu$W) frequencies occur between 10.9 and 36 gigahertz (GHz). They form part of a larger continuous range of frequencies covering 0.225–100 gigahertz, which are usually subdivided into bands designated by letters.

**kbps** KiloBits Per Second. A rate of data transmission of $10^3$ binary digits (BITs) per second (1000 bits per second).

**KC** KiloCycle. A representation previously used to indicate one thousand cycles per second (C/S) now deprecated in favour of kHz.

**KCS** (1) Kilo Characters per Second. One thousand characters per second is a measurement of data transmission speed that enjoys general acceptance.

(2) Kansas City Standard. The standard for cassette tape recording and playback of data on audio cassettes. KCS is so-called because of a meeting of manufacturers and technical editors in Kansas City who decided on data encoding techniques using standard audio cassettes and recorders.

**kg** KiloGram. The SI unit of mass. It is equal to the international prototype of the kilogram that is kept at Sevres in France in the form of a piece of platinum-iridium.

**kHz** KiloHertz. One thousand hertz (Hz).

**KIPS** Kilo Instructions Per Second. A method of rating the speed of a machine, standing for one thousand operations per second. Large-scale integration (LSI) microprocessors (MP) can execute approximately 500 KIPS, whilst some of the larger computers can execute around 3000 KIPS.

**KISS**  Keep It Simple Sir (or Stupid). Jargon used in certain activities of computer programming.

**KIT**  (1) A colloquial expression used to describe a system that is assembled by the user. It does not necessarily imply an assemblance of components or parts, as it is often applied to a marriage of stand-alone elements, or production equipment.
(2) Key Issue Tracking. A store of information held on computer files in the USA, that provides a continually-updated index of current important topics.

**KLIC**  Key Letter In Context. An index system that lists available computer programs or documents. It is similar to keyword in context (KWIC) indexing system but is based on letters rather than words. Permuted lists of terms are produced, sorted by each letter of every term, with the remainder of the term being entered.

**KLU**  Key and Lamp Units. Equipment that can be installed along with a subscriber's telephone, usually in an office environment, to contend with incoming and outgoing calls on a number of exchange lines, extensions or private circuits.

**km**  KiloMetre. One km is equal to $10^3$ metres (m,2) (1000 metres).

**KP**  Start-of-Pulse signal. A term used in the USA to describe a signal that is sent forward down a telephone line to indicate that another signal identifying the party to be called or the onward route of the call is about to be transmitted.

**KPIC**  Key Phrase In Context. An index system that lists available computer programs or documents, similar to keyword in context (KWIC) and keyletter in context (KLIC) indexing systems, but based on phrases rather than words or letters.

**KPO**  Key Punch Operator. A person who controls a device that perforates cards in specific positions. The device works under the guidance of a computer (CMPT) or a keyboard (KB) operator.

**KSAM**  Keyed Sequential Access Method. A group of computer library routines together with a structure that allows the storage of data under assigned names. This in turn allows users to obtain records based on the ordering of key field contents.

**KSR**  Keyboard Send/Receive. A combination teletypewriter (TTY) transmitter and receiver that has transmission capability manually from the keyboard (KB) only, but no means of automatic origination or retention of data.

**KTR**  Keyboard Type Reperforator. A special kind of teletypewriter (TTY) that receives, punches and types on paper tape (PT). It can simultaneously punch and transmit.

**kV** KiloVolt. One kV is equal to $10^3$ volts (V,1) (1000 volts).

**kW** KiloWatt. One kW is equal to $10^3$ watts (W,2) (1000 watts).

**KWAC** Key Word And Context. An index system that lists available computer programs or documents. It is similar to keyword in context (KWIC) indexing system, but each significant word is brought to the beginning in alphabetical order. The title has the keyword first, then the words that follow it, followed by the words that came before the keyword.

**KWIC** Key Word In Context. An index system that lists available computer programs or documents tabulated in alphabetical order of the most informative or significant word in the title of the information. Cf KWOC and *see* KWAC.

**KWOC** Key Word Out of Context. An index that lists available computer programs or documents tabulated with the title used in full under as many significant or informative words from the title of the information as the compiler considers necessary. Compare KWIC.

# L

**Lab PS** LABoratory Power Supply. A laboratory unit that provides a continuously-variable output (O,2) voltage (V,2) over a specified range. A typical unit includes features such as remote sensing, adjustable over-current and over-voltage protection (OVP), as well as power and overload indicators.

**LAMA** Local Automatic Message Accounting. A combination of automatic message accounting equipment and automatic number identification equipment housed in the same office. This American system can automatically process a telephone subscriber's dialled toll call for billing purposes without an operator's assistance.

**LAN** Local Area Network. A computer network operating over a small area linking to computers, electronic mail and word processors (WP) and various other equipment. It can provide communication from site-to-site or provide access to other networks such as the public telephone network and a data transmission network.

**LAPADS** Lightweight Acoustic Processing And Display System. An antisubmarine system with a microprocessor (MP) unit that converts signals received from sonobuoys into digital (D,4) form. It then carries out all the necessary filtering and analysis as well as processing the data for display (D,6).

**LARAM** Line-Addressable Random-Access Memory. A block-access memory device consisting of a number of short charge-coupled device (CCD) shift registers that have common input, output (O,2) and regeneration circuits.

**LARP** Local And Remote Printing. The provision of a computer or word processor (WP) print-out or hard copy near the operator's terminal and if preferred at a more distant point.

**LASER** Light Amplification by Stimulated Emission of Radiation. An amplifier and generator of a narrow coherent beam of electromagnetic energy in the optical or visible light region of the spectrum. Light from the laser is all the same frequency, has a high energy density and can travel great distances without any divergence from its tight beam.

**LASERCOM** Light Amplification by Stimulated Emission of Radiation Computer Output Microfilm. A device that combines LASER and dry film technologies, to enable the laser to mark or write directly onto microfilm. The system results in instant micrographic storage dispensing with any in-between steps.

**LB** Langmuir-Blodgitt. An elegant technique for producing ordered layers or films or organic molecules. Research is towards the development of chemically- and biologically-sensitive semiconductor devices, integrated optics, luminous displays and solar cells.

**L band** L-band microwave ($\mu$W) frequencies occur between 0.39 and 1.55 gigahertz (GHz) and form part of a larger continuous range of frequencies covering 0.225–100 gigahertz which are usually subdivided into bands designated by letters.

**LC** (1) Lower Case. A character that can be printed on paper or displayed on a screen, which is not a capital letter and is not included in the upper case (UC) set of characters.
(2) Line Circuit. Equipment at a telephone exchange that acts to supervize the condition of a subscriber's line, when particular control signals are received.

**L-C** Network containing both inductance and capacitance (C,6). This form of tuned circuit has the product, L-C, of inductance and capacitance constant for a particular frequency.

**LCB** Line Control Block. An area of a memory containing control information for operations on a communication line. It can be divided into several groups of characters treated as a whole (field) most of which can be designated as generalized control blocks.

**LCC** Leadless Chip Carrier. A mounting technique employing a high-density package for very-large-scale integration (VLSI) integrated circuits (ICs).

**LCCC** Leadless Ceramic Chip Carrier. A particular mounting technique employing a high-density package for very-large-scale integration (VLSI) integrated circuits (IC).

**LCCD**  Low Complexity Colour Display. A form of high-quality colour television display providing well-defined alphanumerics and graphics, high-performance displays driven from a logic source, as well as high-quality legibility and sharper colour reproduction.

**LCD**  Liquid Crystal Display. A display that depends on an external light source for visibility and is based on materials that exhibit a regular crystal-like structure even in a liquid state. A normally transparent liquid is sandwiched between two glass plates that have a conductive material on their inner surfaces.

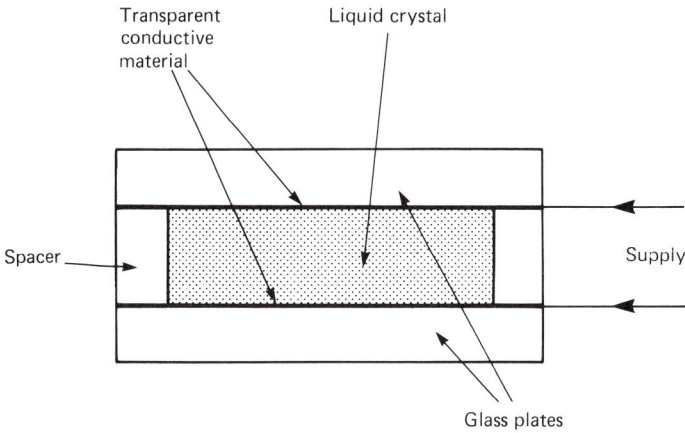

*Figure L.1*  A simplified cross-sectional view of a liquid crystal display (LCD) cell

When subject to an external electric field the liquid becomes opaque and reflective, *see Figure L.1*. As in other displays, characters are normally built up from segments. As they require minimal power, LCDs are used in battery devices such as calculators, digital (D,4) watches and portable digital instruments.

**LCDDS**  Leased Circuit Digital Data Service. A service provided by British Telecom (BT) of high-speed, high-quality microwave ($\mu$W) data transmission links. The communications, between London and 30 other cities, are the first stage of BT's integrated service digital network (ISDN).

**LCL**  LoCaL. A mode of operation in some systems, in which information entered via the keyboard (KB) is displayed (D,6) on a screen but not transmitted on a communication line.

**LD** Laser Diode. A junction diode (D,5) that consists of positive and negative charge carrier regions with a junction region which, when injected with electrons, produces radiation at optical frequencies. In materials like gallium arsenide (GaAs) where emitted light is produced rather than heat, some reflected light trapped by polished ends is eventually emitted to produce laser (LASER) action.

**LDF** LoaD Factor. A ratio that is drawn between the number of units of electrical energy during a given period, and the possible number that would have been supplied had maximum demand been maintained.

**LDRI** Low Data Rate Input. A comparative term relating to the speed with which a device can receive and take in information.

**LDX** Long Distance Xerox. A specific form of communication which combines facsimile (FAX) and xerographic copying.

**LE** (1) Less than/Equal to. A comparison of two values that states whether one value is less than or equal to another. Normally it is used in order for another process or calculation to be carried out. The symbol for LE is ≤.

(2) Local Exchange. A telephone exchange where subscribers' lines are terminated and connected.

**LEA** Longitudinally Excited Atmosphere. A type of laser that uses gas as an active medium and has an electric field that excites the gas in the direction of its flow, i.e. longitudinally.

**LED** Light-Emitting Diode. (a) Widely used as indicators in panels and displays, an LED is a p-n junction (p-n) that emits visible light at relatively low power levels when conducting. These diodes (D,5) with a light-producing characteristic provide light of different wavelengths (colours) by variations in their construction. (b) The symbol LED represents a light-emitting diode on parts lists and schematic diagrams.

**LEO** Lyon's Electronic Office. The first commercial computer (CMPT).

**LET** LETters shift. A function initiated by a letters shift character and performed by a teletypewriter (TTY). It causes the machine to make a physical movement to lower case (LC) so that letters and lower case characters can be printed.

**LEXIS** (1) LEXicography Information Service. A machine-aided translation (MAT) system in West Germany that provides translation between English, French, German and Russian. It consists of an automatic dictionary with a computer output microfilm (COM) facility.

(2) A UK on-line legal information retrieval system of US origin which contains all Statutes, all Statutory Instruments at any time in force, and all reported cases since 1945.

**LF** (1) Line Feed. A specific control character used to advance one printing or display (D,6) position to the next position.
(2) Low Frequency. A frequency between 30 kHz and 300 kHz, forming part of a continuous range of frequencies. LFs are internationally agreed to be radio frequencies in band number 5 and have wavelengths between 1 km and 10 km.

**LFC** Local Form Control. Sited locally, this is a system for off-line data-entry operations by diskette storage of information and unchanging descriptions or specifications (fixed formats).

**LFU** Least Frequency Used. A replacement process or set of rules leading to a desired output (O,2) from a given input, where existing data in a storage area or memory are replaced by new data, the replaced data being the least frequency used items. When it is required to overwrite information already resident in a main memory with new information, the LFU algorithm is used to decide which page or segment of memory is used least.

**LHOTS** Long-Haul Optical Transmission Set. Capable of transmitting lightwave signals for long distances via an optical cable, this type of optical link is used for telecommunications between distribution frames or switching centres.

**LIFO** Last In First Out. (a) A method of queuing where the last item attached to the queue is the first to be withdrawn. (b) A discipline of queueing in a computer where the newest entry in a file is the first to be removed.

**LILO** Last In Last Out. (a) A method of queueing where the last item attached to the queue is the last to be withdrawn. (b) A discipline of queueing in a computer where the newest entry in a file is the last to be removed.

**LinCMOS** LINear Complementary Metal-Oxide Semiconductor. A silicon-gate complementary metal-oxide semiconductor (CMOS) process technology proprietory to Texas Instruments. It combines the low power, low voltage and high input impedance of CMOS integrated circuits (IC) with offsets and voltage (V) swings better than most bipolar devices, and is comparable to some (BIFET) devices.

**LIPL** Linear Information Programming Language. A computer language that enables the user to employ a notation with which he is already familiar.

**LIS** Large Interactive Surface. The automated drafting table employed in plotting and/or digitizing drawings in computer-aided design/computer-aided manufacture (CAD/CAM).

**LISN** Line-Impedance Stabilization Network. A network that simulates the normal impedance of a typical power source impedance. It is used in measuring conducted emissions emanating from equipment to a power line.

**LISP** LISt Processing. A computer language developed to manipulate symbolic strings of repetitive data and used primarily for research rather than for production programming. It is used generally in the varification of mathematical proofs and for developing higher-level languages.

**LLGA** Leadless Land Grid Array. A particular mounting technique employing a high-density package for very-large-scale integration (VLSI) integrated circuits (IC).

**lm** LuMen. Le Système International d'Unités (SI) unit of light flux which is $1/(4\pi)$ of the total light flux that is emitted by a source of 1 candela (CD) intensity.

**lm-hr** LuMen-HouR. The unit quantity of light that equals one lumen of luminous flux flowing for one hour.

**LMS** Level Measuring Set. Instrumentation consisting of a receiving circuit calibrated in decibels (dB), which is used to measure the level of test-signals in telecommunication lines.

**lm-sec** LuMen-SECond. The unit quantity of light that equals one lumen of luminous flux flowing for one second.

**LOC** Large Optical Cavity. A laser diode (LD) in which the material junctions are so arranged to provide a wide optical cavity for light waves to reflect back and forth, to provide the laser action. The cavity has a higher refractive index than the material surrounding it. The device provides an output (O,2) beam and power greater than a light emitting diode (LED).

**LOCAL** Load On CALl. If a particular program is too big to be loaded into the core memory of a central processor unit (CPU), sometimes it can be split into subprograms which can be held on disk or magnetic tape (MAGTAPE) which can be loaded into memory on call.

**LOCAP** LOw-CAPacity. Low-capacity, low-loss paired telecom-munication cable to transmit a pulse-code modulation (PCM) carrier at 6.3 Mbps; used by American companies operating the Bell telephone system.

**LOS** (1) Loss Of Signal. The reduction in amplitude of a transmitted signal such that it can no longer be received.
(2) Line Of Sight. A radio or optical link that has an unobstructed straight-line path between its transmitting and receiving antennae. In practice the line is not straight as atmospheric refraction has to be taken into account.

**LP** (1) Line Printer. A high-speed printer capable of printing an entire line of data simultaneously, 80 to 120 characters long, and then advancing to the next line. These printers normally use rotating drums or chains on which the necessary characters are located. They are normally associated with computers or medium to high-speed terminal services and are much faster than character-at-a-time (*see* CPS) printers.

(2) Linear Programming. A technique employed in mathematics and operational research for determining or solving particular problems involving many variables, where it is required to find the best value or best set of values. The LP technique is not to be confused with computer programming, although problems using the technique can be programmed onto a computer. The mathematical technique used employs a series of linear equations to best allocate any limited resources.

**LPC** Linear Predictive Coding. A technique used for the analysis of speech and its conversion into digital (D,4) coded information.

**LPM** Lines Per Minute. (a) Usually used as a suffix forming an expression indicating the number of lines of text that can be dealt with by a printer or a teletypewriter (TTY). (b) Refers to the speed of operation of a line printer (LP).

**LRC** Longitudinal Redundancy Check. (a) A telecommunications error control based on the formation of a block check that follows preset rules. Information is accumulated at both sending and receiving stations during the transmission of a block. The accumulation is termed the block-check character (BCC) and is transmitted as the last character in the block. A transmitted BCC is compared with an accumulated BCC to determine when a good transmission block has been made. (b) A procedure for checking errors in a computer program.

**LRU** Least Recently Used. A computer processing routine based on the length of time since the last access to a particular page or segment of main memory was made, in order to determine which part of the memory is to be overwritten.

**LSB** Least Significant Bit. The numeral, or binary digit (BIT) in the binary scale of notation, that contributes the smallest quantity to the value of a number and in practice is the right-most digit (DGT).

**LSC** Least Significant Character. The character in the right-most position in a number or word.

**LSD** Least Significant Digit. The significant digit (DGT) that contributes the smallest quantity to the value of a number and in practice is the right-most digit.

**LSI** Large-Scale Integration. The design and production of integrated circuit or microcircuit chips (CHIPs). The solid state circuitry is produced on a single piece of material, normally silicon. LSI denotes circuits of complexity of up to 16 kilobits (KB,2).

**LSS** Loop Switching System. A term used in the USA to

describe a system that accommodates more telephone subscribers than line pairs in the system. It is achieved by concentrating traffic from a remote terminal, onto a smaller number of lines, to a complementary terminal at the local exchange (LE).

**LSTTL** Low-power Schottkey Transistor-Transistor Logic. A widely used and versatile variant of the transistor-transistor logic (TTL) range, with a propagation delay similar to standard TTL (9.5 nanoseconds) but with a much lower power consumption (2 milliwatt).

**LT** (1) Less Than. A comparison of two values that states whether one value is less than another. Normally it is used in order for another process or calculation to be carried out. The symbol for LT is <.

(2) Low Tension. A low voltage (LV) that is used to supply power for the heaters or filaments of an electron tube. Tubes operated from batteries have an LT of 1.4–2 volts (V) and indirectly-heated tubes which usually operate from an alternating current (AC) supply have an LT of 6.3 volts (V).

**LTRS** LeTteRs Shift. (a) A control character that indicates following characters on a teletypewriter (TTY) that are subject to figures shift (FIGS). (b) A physical shift used in a teletypewriter (TTY) that enables alphabetic characters to be printed. (c) The name given to the character that causes the letters shift.

**LUF** Lowest Useful high Frequency. *See* LUHF.

**LUHF** Lowest Useful High Frequency. The lowest specific frequency in the high-frequency (HF) band (i.e. 3–30 MHz) that can be used for communication via the ionosphere at any particular time.

**LV** Low Voltage. Until 1976, LV had stood for voltages not exceeding 250 volts (V). Now European regulations use the term for voltages up to 1000 V between conductors and 600 V between conductor and earth.

**LVA** Large Vertical Aperture. A radar antenna that has a large vertical aperture measuring 1.53 metres (m) as opposed to 0.46 m of a linear array antenna. The LVA antenna contains a vertical array of radiating elements that provide more flexibility in the control of the radar's transmit and receive patterns, thereby providing improved performance.

**LVD** Liquid crystal Visual Display. A low-power-consumption panel that can provide a clear display (D,6) even in bright sunlight. It is a development of a liquid crystal display (LCD) screen that incorporates on a silicon wafer approximately 32 000 dots and a very-light-sensitive large-scale integrated (LSI) circuit.

**LX** LuX. The SI unit of illuminance. It is the quotient of the luminous flux incident on a surface area by that area. It is an absolute unit of illumination equal to one lumen per square metre (m).

# M

$\mu$ Micro. A decimal prefix commonly used in association with a base unit of the SI system of units, indicating a multiplication factor of one millionth, $10^{-6}$ of that unit. For example, one microsecond ($\mu$s) is equal to 0.000 001 second.

$\mu$**A** MicroAmpere. One millionth of an ampere (A), $10^{-6}$ ampere or 0.000 001 ampere.

$\mu$**m** MicroMetre. One millionth of a metre (m), $10^{-6}$ metre or 0.000 001 metre.

$\mu$**s** MicroSecond. A period of time equal to one millionth of a second, $10^{-6}$ second or 0.000 001 second.

$\mu$**W** MicroWave. Electromagnetic wavelength shorter than 300 cm (frequencies above 1 GHz).

**m** (1) Milli. A decimal prefix commonly used in association with a unit of the SI system of units, indicating a multiplication factor of $10^{-3}$ of that unit. For example, one millimetre (mm) is equal to 0.001 metre.

(2)Metre (meter, US). The SI unit of length. It is defined as a length equal to 1 650 763.73 wavelengths in a vacuum of the radiation associated with the transition between the levels $2P_{10}$ and $5d_5$ of the Krypton 86 atom.

**M** Mega. (a) A decimal prefix commonly used in association with a base unit in the SI system of units, indicating a multiplicating factor of $10^6$ of that unit. For example, one megavolt (MV) equals 1 000 000 volts. (b) A prefix used in computing that denotes a multiple of $2^{30}$. One Mbyte or megabyte equals 1 048 576 bytes.

**MAC** (1) Multiple-Access Computer or Multi-Access Computer. A computer system that can receive or despatch input or output (O,2) information from more than one location or terminal.

(2) Machine-Aided Cognition. The application of computers that have artificial intelligence. The system exhibits characteristics commonly connected with human intelligence, including learning, logic, problem solving, reasoning and the ability to understand natural language.

**MACRO** MACROcode or MACROprogram. A computer source-language statement, routine or program that is able to be understood by a computer compiler or processor with or without the assistance of another routine to decode instructions.

**MAGB** Microfilm Association of Great Britain. A group that endeavours to increase the standard of the production of microfilm and promote its efficient use.

**MAGTAPE** MAGnetic TAPE. Available in many forms, magnetic tape normally consists of a plastics substrate on which a magnetic medium is formed from a deposit of an oxide such as ferric oxide ($Fe_2O_3$) chromium. Recent tapes use particles of metal. The tape may be wound onto reels either in a closed *cassette* container with 4 mm tape or an open *reel-to-reel* with 12 or 25 mm tape. Generally cassettes are used for low to medium quality or domestic audio or microcomputer recording while reel-to-reels are used with high-quality audio or large mainframe computers.

**MAI** Machine-Aided Index. The use of a store of information held on a computer file to produce an index for documents. The documents are provided with titles that give an indication of their contents.

**MAMMAX** MAchine-Made and Machine-Aided indeX. A form of machine-aided index (MAI) in which documents in machine readable form are scanned by a computer, and entries are written into the index in a form that complies with instructions in a particular extraction program.

**MANIAC** Mechanical And Numerical Integrator And Computer. One of the early computers (CMPT). *See* ENIAC.

**MAP** (1) Microprocessor Application Project. A project launched in 1978 to further the uses of microelectronics for products and processes in the UK. It was initiated and promoted by the Department of Industry and organized by the National Computing Centre.

(2) Memory MAP is one that shows how computer programs are allocated to memory locations and can also indicate where memory exists in the computer.

(3) Bit (binary digit) MAP indicates if a particular item of information is present or absent (is being used or is unused). The item is represented by a single binary digit (BIT), 0 or 1.

**MAR** Memory Address Register. A special memory storage area, used to hold the identification of the location of a selected word either to be read from or to be written into the main memory.

**MARC** MAchine-Readable Cataloguing. Developed by the British Library for the US Library of Congress, this system was designed for the production of records. The records contain information of description of books, articles, reports and documents that are capable of being read by a computer.

**MARECS** European maritime satellite system. A communication satellite being set up by a provisional organization established by the telecommunications and operating agencies of Europe (Interim Entelsat) to provide telecommunication services for ships at sea.

**MARISAT** MARItime SATellite service. Making use of geostationary satellites over the Pacific and Atlantic oceans, this service enables communications to be established between ships in either ocean.

**MASER** Microwave Amplification by Stimulated Emission of Radiation. A term used to describe a general class of microwave ($\mu$W) amplifiers (AMPs) which are based on molecular interaction electro-magnetic radiation. The maser is non-electronic in its working principle, providing low-value noise emissions, especially if operated at very low temperatures.

**MASS** Multiple Access Sequential Selection. A particular method of storing and retrieving information that is used by a computer.

**MAT** Machine Aided Translation. A system of translation employing a computer-based multilingual lexicon. The system assists a human translator by providing the swift translation of particular words and/or terms, from one spoken human language to another.

**MATV** Master Antenna TeleVision. Similar to a community antenna television (CATV), an MATV is designed to provide television pictures but is on a smaller scale than a community system. A typical example would be an antenna serving a block of flats.

**MB** MegaByte. A million groups of adjacent binary digits (BITs) operating as a unit (BYTEs). It is employed as a measure of the storage capacity of a device.

**MBM** Magnetic Bubble Memory. A memory device employing a very thin layer of magnet garnet set out in its natural state as ribbon-shaped strips or magnetic domains. When an external magnetic field is applied to the material the domains contract into stubby cylinders, which when viewed from the top under a microscope look and behave like bubbles, as shown in *Figure M.1*. When influenced by the external fields the bubbles can be manipulated to represent information binary digits (BITs).

**Mbps** MegaBits Per Second. A rate of transmitting information equal to 1 000 000 binary digits (BITs) per second.

**MCAR** Machine Check Analysis and Recording. A diagnostic procedure initiated by an automatic interruption that occurs

*Figure M.1* Bubble and strip domains in a garnet film, employed in a magnetic bubble memory (MBM) *(courtesy Plessey Microsystems Ltd.)*

126

when the programmed checking of machine functions (machine check) circuits detect a machine error.

**MCB** Miniature Circuit-Breaker. A device that automatically breaks the normal current flow of a circuit under fault conditions such as a short-circuit, it is rated at around 60 amps (A) or less.

**MCH** Machine Check Handler. A device that controls the turning on of a protective device when certain conditions arise within a machine. The machine can be programmed to stop, to run a separate correction routine, or to ignore the condition.

**MCP** (1) Master Control Program. A program that controls all phases of a job. It directs all equipment functions and the flow of all data; provides for automatic error detection and correction; directs the operator with printed instructions and adjusts system operation to changes in system environment.

(2) Message Control Program. A specially designed routine or program used to control the sending or receiving of communications or messages to or from a remote terminal.

**MCR** Master Control Routine. A supervisory routine in a program consisting of a series of subroutines. It controls the linking of the other subroutines and controls the overlaying of the various segments of memory as required.

**MCU** Microprogram Control Unit. Used in bipolar microprocessors (MP), this functional unit maintains and generates microprogram addresses. It is also used to control carry-and-shift operations, and in conjunction with an interrupt (INT) control unit sets up the interrupt structure.

**MCVD** Modified Chemical Vapour Deposition. A modified form of the inside vapour-phase oxidation (IVPO) process used for the production of optical fibres (OF). Burners travel along the glass tube and soot particles are created inside rather than being deposited by the heat flame on the outer surface, as in the outside vapour phase oxidation (OVPO) process.

**MCVF** Multi-Channel Voice Frequency. A multiplexing (MUX) method for transmitting a number of signals over one single telephone circuit used by British Telecom (BT).

**MCW** Modulated Continuous Wave. A particular carrier wave that is modulated by a steady audio frequency (AF) tone.

**MDF** Main Distribution Frame. A distribution frame used in telecommunications that has connections on either side. The line or lines external to the telephone exchange are connected on one side and the internal cables of the exchange equipment are connected on the other.

**MDR** Memory Date Register. A special memory storage area that holds the last data word read from or written into the memory location, addressed by the contents of the memory address register (MAR).

**MDS** (1) Microprocessor Development System. Essentially a system for producing software for microprocessors. A computer-controlled system which employs various devices or peripherals to aid the application development of microprocessors (MP). *Figure M.2* shows what is generally accepted as the configuration of an MDS.

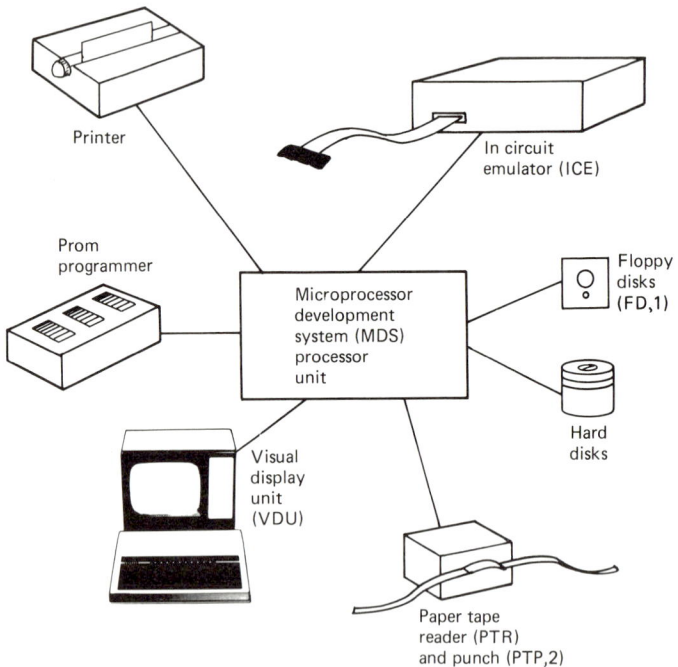

*Figure M.2* Typical configuration of a microprocessor development system (MDS)

(2) Minimum Discernible Signal. The least or smallest value of input signal to any device or circuit that is just able to produce a discernible change in the output (O,2).

**MDT** Mean Down Time. The average period during which a computer is malfunctioning or not operating correctly due to machine failures.

**MECL** Motorola Emitter Coupled Logic. A family of emitter coupled logic (ECL) devices manufactured by Motorola who, along with Fairchild, were the originators of ECL when they were introduced in the early 1960s.

**MELF** Metal Electrode Face-bonding. The name given to a group of leadless electronic components which have metal tips as their terminations, as opposed to leads or legs. The tips are then employed to attach the device directly to the tracks on the printed circuit side of a printed circuit board (PCB).

**MESFET** MEtal Semiconductor Field-Effect Transistor. A junction field-effect transistor (JFET) that has a barrier formed by a junction between metal and semiconductor (a Schottky barrier) as a gate electrode rather than a semiconductor junction.

**METALANGUAGE** An artificial computer language employed to define another computer language of which Backus-Naur Form (BNF) is an example.

**METAPLAN** Methods of Extracting Text Automatically Programming LANguage. A computer programming language used for the retrieval of required portions of text.

**METASYMBOL** METAlanguage SYMBOL. Any of the symbols employed in the METALANGUAGE artificial computer language.

**MF** Medium Frequency. Frequencies between 300 to 3000 kHz forming part of a larger continuous range of frequencies. MFs are considered by international agreement to be radio frequencies in band number 6 and have wavelengths of between 100 m and 1000 m.

**MFLOP** Million FLOating Point. A measure of computing power that refers to the number of operations per second made when the position of the decimal point of stored numbers changes during computing.

**MFSK** Multiple-Frequency-Shift Keying. A form of communication where a separate frequency is used to transmit a digital data input code. Ten discrete frequencies are used for the ten numerals of the decimal system of notation. *See also* FSK, frequency shift keying.

**MFSS** Multi-Frequency Signalling System. A system that transfers information using alternating currents (AC) with frequencies in the telephone speech band, to set up, control and release calls. The information is represented by compound signals, made up from a number of voice-frequency currents chosen from a set of available currents.

**MFT** Multi-programming with a Fixed number of Tasks. The concurrent execution of two or more programs within a single computer, to accomplish a set number of units of work.

**MH** Manual Hold. When a call is made via the operator in an automatic telephone system, MH is established when the telephone caller's line is held in the busy state until the operator releases it.

**MHD** MagnetoHydroDynamic. The physical behaviour of electrically conducting fluids (gases or liquids) in the presence of electric or magnetic fields. It can also relate to the observation and study of these phenomena.

**MIC** Microwave Integrated Circuits. A diverse range of fabrication techniques which range from microstrip circuits through various thin-film technologies (TFT,2) employing glass and alumina substrates, up to monolithic microwave integrated circuits (MMIC) employing bulk silicon bipolar, silicon-on-sapphire (SOS) and gallium arsenide (GaAs) processes. At present the majority of MIC manufacturers produce hybrid devices utilizing a mixture of thin-film and stripline techniques, achieving such circuit functions as oscillators, mixers and amplifiers in combinations to suit many applications and for frequencies up to 60 GHz or more.

**MICC** Mineral-Insulated Copper-Covered. A type of electrical cable that consists of conductors insulated from one another and the exterior by a mineral material, having an overall sheathing of copper. The cable is often used in harsh or damp environments, employing tight fitting seals (glands) at either end upon entering and leaving enclosures.

**MICR** (1) Magnetic Ink Character Reader. A device capable of converting into a machine code characters which are printed in magnetic ink and are readable by humans.
(2) Magnetic Ink Character Recognition. An encoding system, employing magnetic ink, used by banks to enable cheques to be handled automatically.

**micro** (1) A colloquial expression often used as an abbreviation for a microcomputer or a microprocessor.
(2) Numerical prefix, *see* $\mu$.

**MICRO** Multiple Indexing and Console Retrieval Operation. Interacting directly with the central processor unit (CPU) of a computer, a MICRO is designed to retrieve, rank and qualify document entries in an index.

**MIIS** Metal Insulation Insulation Semiconductor. A sandwich of semiconductor materials that employs stored charges between two layers of insulation to produce a nonvolatile memory. When used with the silicon-on-sapphire (SOS) process a memory is produced that is independent of the external power supply.

**MIKE** MICrophone. A transducer that converts sound energy (acoustic vibration) into electrical energy (signals). It is the initial device used in telephones, in broadcast transmission and all forms of electrical sound recording. Types include carbon, crystal, moving-coil, capacitor (C,6) and ribbon types. *See Figure M.3.*

130

**MIL** (1) A unit of length equal to one thousandth of an inch.
(2) MILitary. Usually refers to components, devices or systems produced specifically for or to meet stringent military standards and specifications. Normally written as a prefix, for example MIL-SPEC for 'military specification' or MIL-STD for 'military standard'. The US Defense Electronic Supply Centre military specification body is responsible for co-ordinating MIL specifications and standards.

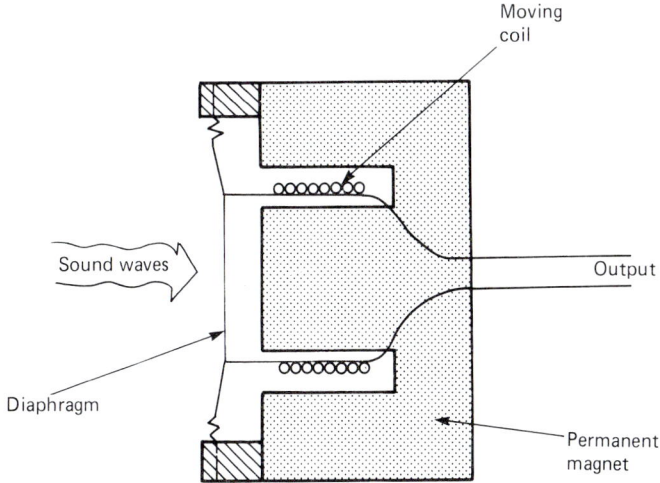

*Figure M.3* A cross-sectional view of a moving-coil microphone (MIKE)

**MILEPOST** Middlesborough Initiative Local Electronic Provision Of Service Technique. Originally an initiative within the area of Middlesborough to take basic National Physical Laboratory Scrapbook Software and add to it additional services in order to produce a more comprehensive computer processing service. It was further developed and launched by British Telecom (BT) in 1983 as a business system. The interactive facilities it offers include information storage and retrieval, word processing (WP), electronic mail, multi-activity project management and management information.

**MILL** Colloquial expression for a central processor unit (CPU) used in some computer systems. A CPU grinds or crunches information as a mill grinds wheat.

**milli** A prefix denoting one thousandth part of the unit which follows or is attached to it. For example, millisecond is equal to one thousandth of a second. *See* m(1).

**MIL-STD-883** MILitary STanDard. The basic US military

standard for the reliability of semiconductors. It has three classes: A for aerospace, B for avionics and C for ground use. *See* MIL 2.

**MIMD** Multiple Instruction stream, Multiple Data stream. A particular construction or architecture of computers. Multi-processor systems are MIMDs.

**MIN** MINimum. The lowest or least possible size or quantity.

**MINIC** MINimal Input Cataloguing. A particular system employed for the efficient cataloguing of documents.

**MIPS** Millions of Instructions Per Second. A comparative measure of computing power, the number of commands that can be processed per second. 1 MIPS would describe a moderately powerful minicomputer, 20 MIPS would be a medium sized computer and 200 MIPS would be an array processor.

**MIS** (1) Metal Insulator Semiconductor. A device consisting of a metal top contact layer, a second layer of very thin insulating material and a bottom layer of p-doped (p-type) semiconductor material. An example is a solar cell constructed in a sandwich. (2) Management Information System. (a) A type of data processing (DP) system designed to provide management and other specified supervisory personnel with information of a desired nature. (b) A process of communication where data is processed and recorded and any problems are flagged in order that higher-level decisions can be made, and also so that management can review the overall progress of a project.

**MISFEED** MISaligned FEED. The failure of a punched card machine (PCM) to feed a card from its hopper; often caused by damaged punched cards.

**MISFET** Metal Insulator Silicon Field-Effect Transistor. A field-effect transistor (FET) constructed from a wafer of semiconductor material. The wafer has two highly doped regions of opposite electrical polarity diffused into it, forming source and drain regions. An insulating layer is also formed on the surface between these regions and a conductor is deposited on top of this layer to form a gate electrode.

**MIST** Metal Insulator Silicon field-effect Transistor. *See* MISFET.

**MIT** Master Instruction Tape. A tape, most often a magnetic tape (MAGTAPE), onto which all the programs for a system of various computer runs are recorded. It is considered to be a fundamental part of an operating system (OS).

**MIVPO** Modified Inside Vapour Phase Oxidation. A modified form of the inside vapour-phase oxidation (IVPO) process. *See* (MCVD), modified chemical vapour deposition.

**MKR** MarKeR. Apparatus that controls, selects and sets up telephone traffic paths through a switching stage of a switching centre.

**MKS** Metre, Kilogram and Second. A system of absolute units where the fundamental units of length, mass and time are the metre, kilogram and second respectively. There is also a fourth fundamental quantity for the permeability of free space. The MKS system was used as the basis of the SI system of units.

**MLB** MultiLayer Board. A special type of printed circuit board (PCB), with two or more layers of circuit tracks, which is able to increase logic speed and packing density, whilst cutting electrical crosstalk.

**MLC** Multi-Line Controller. A hardware unit that can be added to an advanced research project agency network (ARPANET) packet-switched computer network to provide the facility to handle up to 64 different terminals at varying signalling speeds (baud rate – *see* BAUD).

**mm** MilliMetre. One thousandth of a metre (m).

**MMF** MagnetoMotive Force. The magnetic analogy of electromotive force (EMF). It is the force that gives rise to magnetic fields. Symbol Fm.

**MMIC** Monolithic Microwave Integrated Circuit. A microwave integrated circuit (MIC) that employs bulk silicon bipolar, silicon-on-sapphire (SOS), and gallium arsenide (GaAs) processes.

**MMS** Microfiche Management System. A system employed for the storage and retrieval of information held in microfiche. Similar to computer output microfilm (COM).

**Mn** MaNganese. A grey-pink, hard and brittle metal, atomic number 25, that is often used as an alloying element. It is also used in some forms of primary batteries.

**MNOS** Metal-Nitride-Oxide Silicon (or Semiconductor). An electrically alterable read-only memory (ROM) in which each individual memory cell consists of a metal-oxide semiconductor (MOS) transistor with an overlaying signal gate electrode, as shown in *Figure M.4*. Some ROMs are preprogrammed by the manufacturer, whereas others are programmed by the user by burning-out selected fusible links (FL) prior to installation. In MNOS structures charge is stored at the interface between silicon nitride and silicon oxide to provide the long-term memory action.

**Mo** Molybdenum. A metallic element atomic number 42, atomic weight 95.94, melting point 2625°C, boiling point 4800°C and resistivity 4.77 microhm $cm^2$. It has physical properties similar to those of iron and is used in wire form for the filaments of vacuum tubes, for electrodes of mercury-vapour lamps and for winding electric resistance furnaces.

**MOBL** Marco Orientated Business Language. A specific computer programming language that permits users to employ a notation with which they are already familiar.

**MOCVD** Metallo-Organic Chemical Vapour Deposition. A method for depositing very thin and ultra-pure layers of one material onto a substrate. It is well established for the production of gallium arsenide (GaAS) and can be used in the indium phosphide coating process.

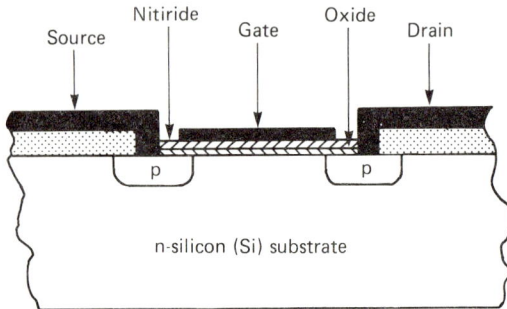

*Figure M.4* A cross-sectional view of a unipolar metal-nitride-oxide silicon (MNOS) structure

**MOD** MOving Domain. A memory device that is non-volatile like a core memory but much smaller, and is reasonably fast in operation. Unlike a bubble memory it requires no magnetic bias to form and hold the domains and can be batch-fabricated using conventional microcircuit techniques.

**MOD/DEMOD** MODulating/DEMODulating. *See* MODEM.

**MODEM** MOdulator/DEModulator. A unit that modulates and transmits communications signals, and demodulates a signal that is received at a data station or terminal. The term can also be applied to units carrying out other functions like multiplexing (MUX). *See Figure M.5.*

**mol** MOLe. The SI unit of amount of substance. It is the amount of chemical substance that weighs its own molecular weight in grams.

**MOL** (1) Multiple On-Line programming. A facility providing interactive program development to a number of user terminals simultaneously.

(2) Machine-Orientated Language. A method of notating information that is intelligible to a computer or specific machine. These languages often include information that directly controls machine operations.

134

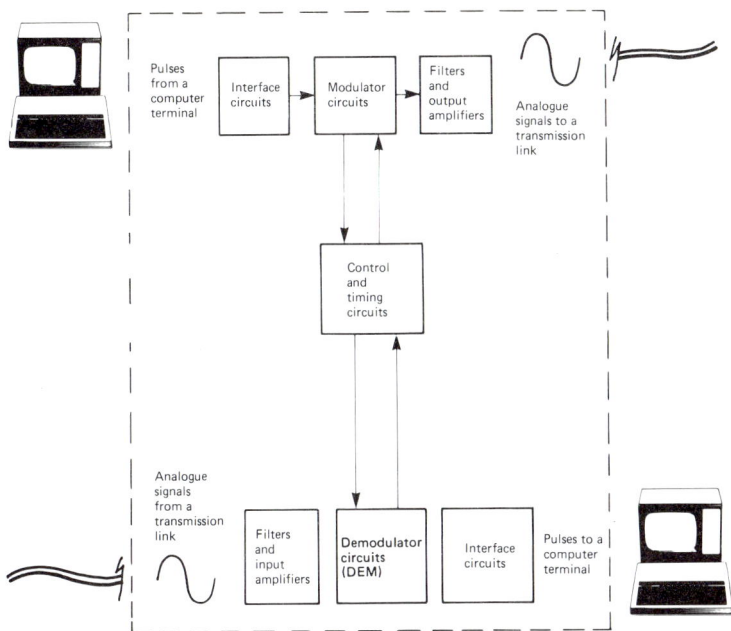

*Figure M.5*   Typical block configuration of a modulator-demodulator (MODEM)

**MOLE ELECTRONICS**   MOLEcular ELECTRONICS. Various techniques employed in the growing of solid-state crystals to form transistors, diodes (D,5) and resistors (R,2) in one mass for microminiaturization.

**MOP**   Multiple On-line Programming. An interactive set of computer software for the overall control of the operation of a computer system used on International Computer Laboratories (ICL) systems.

**MOS**   Metal-Oxide Semiconductor. The three specific layers of material which are used to form the gate structure of a field-effect transistor (FET). MOS devices tend to have a high capacitance (C,5) n-doped (n-type) silicon substrate which makes them slower than bipolar devices, but they achieve higher functional density (D,3) with fewer process steps which reduces fabrication costs.

**MOS DMOS**   Double-diffused Metal-Oxide Semiconductor. A metal-oxide semiconductor (MOS) device that has had a second diffusion process. This results in a product able to handle higher voltages (V) and currents (I) whilst experiencing very little parasitic capacitance (C,5) and low noise. *See* DMOS.

135

**MOSFET** Metal-Oxide Silicon Field-Effect Transistor. A field-effect transistor (FET) constructed from a wafer of semiconductor material with two highly doped regions of opposite electrical polarity diffused into it forming source and drain regions. An insulating layer of silicon oxide is also formed on the surface between these regions and a conductor of aluminium or very highly doped polysilicon is deposited on top of this layer to form a gate electrode. *See Figure M.6.*

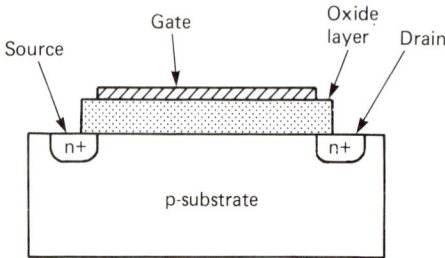

*Figure M.6* A simplified cross-sectional view of an n-channel metal-oxide silicon field-effect transistor (MOSFET)

**MOSRAM** Metal-Oxide Semiconductor Random-Access Memory. A random-access memory (RAM) storage medium that employs metal-oxide semiconductor (MOS) transistor cells to store binary 1s and 0s.

**MOSROM** Metal-Oxide Semiconductor Read-Only Memory. A read-only memory (ROM) storage medium that employs metal-oxide semiconductor (MOS) transistor cells to store binary 1s and 0s.

**MOST** Metal-Oxide Silicon Transistor. A field-effect transistor (FET) that employs a sandwich of metal and oxide layers. *See* MOSFET.

**MP** MicroProcessor. Containing all the essential elements of a central processor unit (CPU), this integrated circuit (IC) includes the control logic, instruction decoding and arithmetic processing circuitry. But in order to be useful the microprocessor chip (CHIP) or combination of chips is married with a memory and input/output (I/O) circuit chips to produce a microcomputer.

**MPP** MolyPermalloy Powder. Powder used to manufacture high flex magnetic powder cores consisting mainly of nickel.

**MPS** MultiProgramming System. Some computers provide a technique for handling the concurrent execution of a number of routines or programs. The system accomplishes this by overlapping or interleaving their execution, so that one program is executed during the input/output (I/O) waiting time of another program.

**MPU** MicroProcessor (or MicroProcessing) Unit. The main hardware constituent of a microcomputer. Normally assembled on a printed circuit board (PCB) the MPU is not housed in an enclosure and does not have a power supply. It consists of a microprocessor (MP), a main memory, input/output (I/O) interface devices and a clock (CLK) circuit. There are also a buffer, drive circuits and various passive circuit elements.

**MPX** MultiPlex. The technique of transmitting two or more messages simultaneously using the same channel (CHNL,(a)) or transmission medium. This can be achieved using time division multiplexing (TDM) or frequency division multiplexing (FDM), the two most commonly used techniques.

**MPY** MultiPlY. An operation performed in computing by two binary digits (BITs) whereby the result is 1 (true), if and only if both digits are 1 (true). If either or both digits are 0 (false), the result is 0 (false). Also known as AND.

**MRDF** Machine-Readable Data File. A file of information that is capable of being read by a computer.

**ms** MilliSecond. One thousandth of a second, 0.001 of a second or $10^{-3}$ second.

**MS** Molecular Stuffing. A clearly defined and ordered process for producing graded refractive index optical fibres (OF). It uses five main steps which are glass melting, phase separation, leaching, dopant introduction and consolidation.

**MSB** Most Significant Bit. The numeral or binary digit (BIT) in the binary scale of notation that contributes the largest quantity to the value of a number and in practice is the left-most digit.

**MSC** (1) Most Significant Character. The character in the left-most position in a number or word. (2) Main Switching Centre. A fully interconnected switching centre situated in a four-wire switched trunk transit network.

**MSD** Most Significant Digit. The significant digit (DGT) that contributes the largest quantity to the value of a number and in practice is the left-most digit.

**MSI** Medium-Scale Integration. The design and production of integrated circuits (ICs) or microcircuits (CHIPs) The solid state circuitry is produced on a single piece of material, normally silicon. MSI denotes circuits of medium complexity, i.e. fewer components in a given area than large scale integration (LSI).

**MT** (1) Magnetic Tape. Available in many forms, magnetic tape normally consists of a plastics substrate on which a magnetic medium is formed from a deposit of an oxide such as ferric oxide ($Fe_2O_3$) or chromium. Recent tapes use particles of metal. The tape may be wound onto reels either in a closed *cassette* container with 4 mm tape or an open *reel-to-reel* with 12 or 25 mm tape. Generally cassettes are used for low to medium quality or domestic audio or microcomputer recording whereas reel-to-reels are used with high quality audio or large mainframe computers.

(2) Mechanical Translation. The automatic translation from one code, computer language or other system or representation to another code, computer language or system of representation.

(3) Machine Translation. Human language translation performed by computers or similar equipment.

**MTBF** Mean Time Between Failure. A measure of the reliability of equipment. If the particular unit can be repaired MTBF is the average time between failure; if the equipment cannot be repaired it is the time until the first failure.

**MTBM** Mean Time Between Maintenance. A measure of the performance of equipment. It provides a guide to how well hardware performs, without necessarily failing completely.

**MTF** Mean Time to Failure. The average length of time for which a particular system, device, component, or any part or sub-system that can be separately tested, continues to work without failure.

**MTL** Merged Transistor Logic. A form of logic that employs two multi-collector transistors. It evolved from diode transistor logic (DTL), a family of bipolar integrated circuits (BIC) which are very compact and provide high speed operation along with a functional packing density (D,1). MTLs use less power, are more stable and cheaper, can be easily mass produced and are two orders of magnitude faster than transistor transistor logic (TTL). Also known as integrated injection logic (IIL or $I^2L$).

**MTST** Magnetic Tape Selectric Typewriter. A typewriter that is able to provide some text handling and editing facilities, but is not as complex as a word processor.

**MTTR** Mean Time To Repair. A measure of the reliability of equipment or a system, as opposed to mean time between failure (MTBF).

**MTU** Magnetic Tape Unit. A piece of equipment that records electrical signals on a magnetic tape (MAGTAPE or MT), using the tape as a storage medium. The tape, which is flexible, is wound on reels and passes a read/write head. High tape speeds can be used as there is no physical contact between the tape and the read/write head.

**MUF** Maximum Useable Frequency. The highest possible

frequency used for radio communication between terrestrial points at one time, via the ionosphere. The MUF is usually the optimum frequency to use for long-distance (DX,3) communications.

**MULDEM**  MULtiplexor-DEMultiplexor (MULtiplexer-DEMultiplexer, US). Analogous to a modulator/demodulator (MODEM), an item of equipment that contains both a digital multiplexor (MUX,2) and a digital demultiplexor (DEMUX,2).

**MULDEX**  MULtiplexor-DEmultiplexor. *See* MULDEM.

**MULTICS**  MULTIplexed Information and Computing Service. A multi-user computer system, developed by the Massachusetts Institute of Technology, that is now commercially used.

**MUSA**  (1) Multiple Unit Steerable Aerial. A directive antenna that can be altered or steered to its direction of maximum sensitivity by adjusting the phase relationship between the units from which it is assembled.
(2) Multiple Unit Steerable Array. A line of identical rhombic antennas, intended for communications reception, which have their outputs (O,2) combined to provide steerable properties to the array.

**MUX**  (1) MUltiplex. (a) The process of transferring information using a single or common channel (CHNL(a)) or path for the concurrent transmission of more than one information stream. *See* frequency division multiplexing (FDM) and time division multiplexing (TDM). (b) The physical use of one channel (CHNL(a)) or path for the communication of two or more source and sink pairs. *See* space division multiplex (SDM).
(2) MUltipleXor (MUltipleXer, US). A device or technique (often a specialized computer having stored program capability) used to overlap or interleave two or more signals for retransmission at a higher speed. The shared resource for example could be a memory or a bus.

**MV**  Medium Voltage. Now not employed in European standards, MV was a term used to imply a voltage between 250 and 650 volts (V).

**MVT**  Multi-programming with a Variable number of Tasks. The concurrent execution of two or more programs within a single computer, to accomplish an indeterminable number of units of work.

**Mx**  MaXwell. The unit of flux in the now obsolete centimetre, gram, second (CGS) system of units. One maxwell is equal to $10^{-8}$ weber (Wb).

# N

**n**  Nano. A decimal prefix commonly used in association with a unit of the SI system of units, indicating a multiplication factor of $10^9$ of that unit. One nanosecond (ns) is equal to 0.000 000 001 seconds.

**N**  (1) Newton. The SI unit of force. It is defined as the force required to provide a mass of one kilogram (Kg) with an acceleration of one metre (m) per second per second. The unit also relates directly to electrical units as it is the force which, when expended over a distance of one metre, does an amount of work equal to one joule (J).
(2) Negative of sign flag. Usually the most significant bit (MSB), this denotes the sign of a word in two's complement notation.

**NA**  Numerical Aperture. An expression of the light-gathering ability of an optical fibre (OF). It is the sine of the angle of acceptance, θ, which lies between the optical axis of the fibre and the most diverse ray of light that it can accept.

**NAG**  Numerical Algorithms Group. A body which has provided a collection of library routines for a number of numerical functions for use on a variety of computers.

**NAK**  Negative AcKnowledge. The opposite to acknowledge (ACK). In the method of error control that relies on repeating any message with errors, the return signal that reports the error is NAK.

**NAM**  Network Access Machine. A computer programmed to generate the specific procedures necessary to gain access to a network that connects a number of computers, and to enable the user to interact with the network.

**NAND**  Not AND. An AND gate that has a negative output (O,2). It is a logic element that has one output and two or more inputs. The output will stand at its defined logic O state when and if all the inputs are standing at their defined logic 1 states. *See Figure N.1.*

**NASA**  National Aeronautical and Space Administration. One of the bodies that determines standards on behalf of the USA Standards Institute (USASI), covering areas of aeronautics, space science and the related technologies.

**NASORD**  Not in Sequential ORDer. A reference made in computer programming to a file not in sequential order.

**NBFM**  Narrow-Band Frequency Modulation. A frequency modulation (FM) form in which the frequency change is kept at such a level that the signal occupies the same bandwidth as an amplitude modulated (AM) signal carrying the same information.

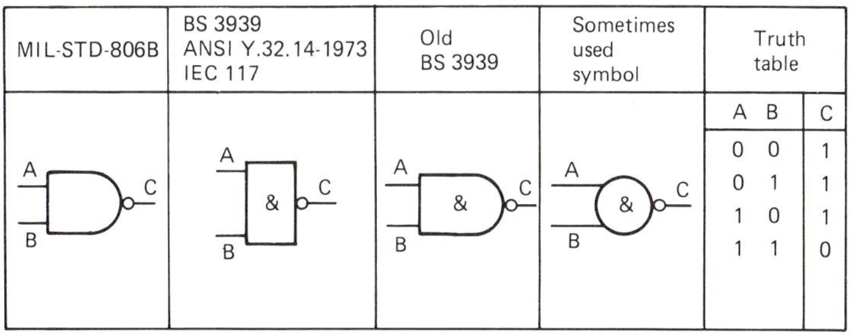

| MIL-STD-806B | BS 3939 ANSI Y.32.14-1973 IEC 117 | Old BS 3939 | Sometimes used symbol | Truth table | | |
|---|---|---|---|---|---|---|
| | | | | A | B | C |
| | | | | 0 | 0 | 1 |
| | | | | 0 | 1 | 1 |
| | | | | 1 | 0 | 1 |
| | | | | 1 | 1 | 0 |

*Figure N.1* The NAND gate. In practice the MIL-STD-806B symbol is the one most likely to be met, although the other logic symbols may be encountered

**NBS** National Bureau of Standards. A body that determines standards on behalf of the American National Standards Institute (ANSI), formerly the USA Standards Institute (USASI).

**NC** (1) Numerical Control. *See* N/C.

(2) No Circuit. A signal used in telephone communications that consists of a low tone (at a frequency of 140 Hz) interrupted at 120 impulses per minute, which indicates to the user that there is no circuit available.

(3) Normally Closed. A switch in which the contacts are closed (contacting) without any external force acting upon it.

**N/C** Numerical Control. The operation of machines by use of punched paper or plastics tape with magnetic spots used to feed digital (D,4) instructions to a machine. The N/C tapes are developed from computer programs.

(3) Normally Closed. A switch in which the contacts are closed (contacting) when no external forces act upon the switch.

**NCC** (1) Network Control Centre. A location in a computer network that controls and in part monitors operations.

(2) Naturally Commutated Cycloconverter. A specialized and restricted AC-to-AC converter having a nonlinear modulating function. It was the first of any type ever to find practical application.

(3) National Computing Conference. Organized by the American Federation of Information Processing Societies, this is an annual conference that takes place in the USA.

**NCP** Network Control Program. A specific computer program that allows process-to-process communication over a computer network.

**NDB** Non-Directional Beacons. Specially designed inland beacons used in conjunction with airborne automatic direction finders (ADF). They radiate a signal that exhibits a sharp reduction in strength directly overhead due to the vertical polarization of transmitting and receive antennas, the effect being employed to obtain a fixed position.

**NDPS** National Data Processing System. A data processing (DP,1) and transmission service run by British Telecom (BT,1). It provides consultancy to commercial customers in the areas of business information systems, computer output microfilm (COM), information retrieval systems and Prestel.

**NDR** Non-Destructive Read. A process of reading information in a memory location without erasing or altering that information in any way.

**NDRO** Non-Destructive Read Out. The sensing of information contained in the internal storage of a computer and the transmission of that information to an external storage unit which inherently does not destroy or erase the record of data that has been read. Reading is usually achieved on tapes, drums, disks and other similar devices, as hard copy on paper or as a display (D,6) on a cathode-ray tube (CRT).

**Ne** NEon. (a) An inert gas that has the atomic number 10. When ionized it exhibits a characteristic red glow and is used in gas discharge tubes also known as 'glow lamps'. (b) A term used to describe very small glow lamps used in electrical/electronic circuits and devices as indicators. (c) A term used to describe very large gas discharge tubes used for illuminated signs or advertizing hoardings.

**NE** Not Equal to. A comparison of two values that states whether one value is unequal or not equal to another. The expression can be used as a direct comparison but is often used so that another process or calculation can be carried out. The symbol for NE is $\neq$ or $<>$.

**NEMP** Nuclear ElectroMagnetic Pulse. A more correct nomenclature for electromagnetic pulse (EMP), being the very high energy pulse generated by the explosion of a nuclear weapon in the atmosphere.

**NEP** Noise Equivalent Power. A measurement of the minimum detectable signal that can be found amongst noise in detection and multiplication processes. It can be defined as the amount of light falling on a photodetector, required to produce an output (O,2) signal equal to the noise output. It is usually expressed in nanowatts.

**NEPHIS** NEsted PHrase-Indexing System. An automated system of indexing subjects listing all the words in the title of a particular document. It is arranged so that every word appears in turn.

**NEPR** Nato Electronic Parts Recommendations. International North Atlantic Treaty Organization (NATO) specifications are produced as Nato electronics parts recommendations.

**NEQ**  Not EQual to. *See* NE.

**NEXIS**  Newspaper database, Mead Data Control. (US). A store of information on computer files providing the full text of newspapers and other agency news. It can be searched directly from a terminal.

**NEXT**  Near-End CROSSTalk. An unwanted energy transfer from a disturbing circuit to a disturbed circuit, where the energy travels in a direction different from that of the signal in the disturbing circuit.

**NFB**  Negative FeedBack. A signal that is returned to an earlier stage of an amplifier (AMP) opposing the input signal at that point. NFB reduces gain in the return or feedback loop, improves linearity and provides more level frequency response. The feedback circuit is built from passive components which are independent of active devices in the feedback loop. It is also known as 'degenerative feedback'.

**Ni**  NIckel. A metal that has the atomic number 28. It is widely used in electronics, in magnetic alloys due to its strong ferromagnetism, and in electrolytic cells as a conductor. *See* Ni-Cad.

**NI**  NonInhibit. Noninhibit interrupts are provided in some computer systems and they include powerfail, auto restart, TTY break, Memory Parity and Protect, Interrupt Program Time Out, and nonimplement Op code.

**NIBBLE**  A colloquial term to describe a 4-binary digit (BIT) word in contrast to a byte, which is generally considered to be an 8-binary digit (BIT) word.

**Ni-Cad**  NIckel-CADmium cell. A secondary cell having potassium hydroxide as an electrolyte, a positive plate (anode) of nickel-oxide/nickel-hydrate mixture, and a negative plate (cathode) or cadmium. It can provide a high rate of discharge along with a steadily maintained voltage (V), properties brought about by the use of cadmium. The cadmium gives the cell a lower charging voltage than a nickel-iron cell (Ni-Fe) and reduces internal resistance.

**NICHROME**  NIckel-CHROMium alloy. A trade name given to an alloy of approximately 63% nickel, 14% chromium and 23% iron. It is used widely for the manufacture of thin-film resistors, wire-wound resistors, and heating elements, due to its ability to operate at very high temperatures and its high resistivity.

**Ni-Fe**  Nickel-FErrous (iron) cell. A secondary cell capable of delivering heavy currents using potassium hydroxide as its electrolyte. The cell has a positive plate (anode) made from a nickel-oxide/nickel-hydrate mixture and the negative plate (cathode) is of iron oxide. The cell is lighter, more durable and

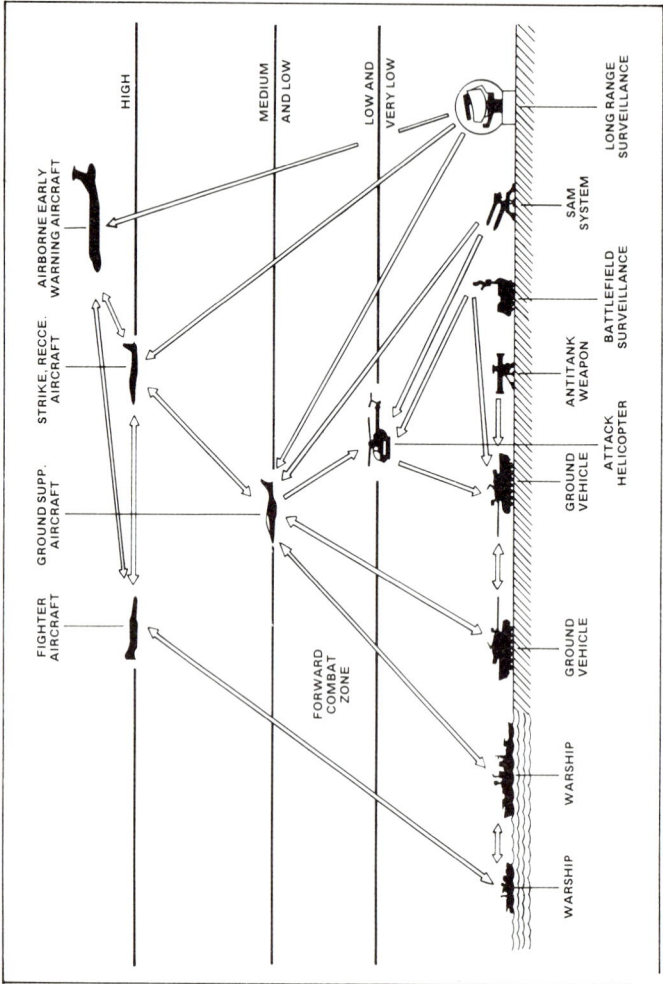

Figure N.2 The NATO identification system (NIS) overall scene (*courtesy* Cossor Electronics Ltd.)

Labels within figure:

HIGH

MEDIUM AND LOW

LOW AND VERY LOW

AIRBORNE EARLY WARNING AIRCRAFT

STRIKE, RECCE. AIRCRAFT

GROUND SUPP. AIRCRAFT

FIGHTER AIRCRAFT

FORWARD COMBAT ZONE

SAM SYSTEM

BATTLEFIELD SURVEILLANCE

ANTITANK WEAPON

ATTACK HELICOPTER

GROUND VEHICLE

GROUND VEHICLE

WARSHIP

WARSHIP

LONG RANGE SURVEILLANCE

144

can operate at lower temperatures ($-30°C$) than a lead-acid cell, but is more expensive and has an electromotive force (EMF) of only 1.2/1.3 volts (V).

**NIP**   Nucleus Initialization Program. Part of an operating system (OS) that resides in main memory or storage, which initializes the resident control program.

**NIS**   NATO Identification System. System embracing whole military scene to provide NATO's identification friend or foe (IFF). It provides fully interoperable identification facilities between all NATO forces on land and sea and in the air, as shown in *Figure N.2*.

**NL**   New Line character. A functional character that controls or demands action of a printer or display, by causing the printing or display position to be moved to the next printing or display line. It combines the functions of a carriage return (CR) and a line feed (LF).

**NLQ**   Near Letter Quality. Dot matrix printers produce characters by forming dots closely together. In general the closer together the dots the better the reproduction (the more it is like typewritten characters). Earlier dot matrix printers had to run slowly for good letter quality, but now high-speed printers are being produced with higher quality characters. They are termed NLQ machines.

**nm**   NanoMeter. One thousandth of a millionth of a metre (m), that is 0.000 000 001 or $10^{-9}$ metres. It is a convenient unit for designating the wavelength of light. It is also equal to 10 angstroms (Å) and is synonymous with millimicron.

**NMC**   Network Measurement Centre. A location on an advanced research projects agency network (ARPANET) packet-switched computer network used to measure performance, normally by means of relatively long-term test.

**NMI**   Non-Maskable Interrupt. A signal sent from an input/output (I/O) device or chip (CHIP) to a microprocessor unit (MPU) to request a service (an interrupt) (INT) at the highest priority that is not affected by an interrupt mask. It is typically used in the event of a power failure.

**NMOS**   N-type or channel Metal-Oxide Semiconductor. Faster in operation than p-channel metal-oxide semiconductor (PMOS) devices NMOS are the newer variety of large scale integration (LSI) chips (CHIP). They are transistor-transistor logic (TTL) compatible and use one power supply and have gates about half the size of PMOS gates.

**NMOS RAM**   N-channel Metal-Oxide Semiconductor Random-Access Memory. A RAM that uses NMOS technology for high speed and low power. They compete with bipolar units for many high-speed cache applications. Housed in a standard 16-pin package and requiring a single 5-volt (V) supply, a typical NMOS RAM has an access time of less than 100 nanoseconds (ns), a maximum power dissipation of less than 600 mW and fully static operation.

**NO** Normally Open. A switch in which the contacts are open (separated) when no external forces act upon the switch.

**NOOP** NO-OPeration. *See* NOP.

**NOP** No-OPeration. An instruction that is used to force a delay of one instruction cycle without altering in any way the status flags or the contents of a memory location.

**NOR** Not OR. An OR gate that has a negative output (O,2). It is a logic element that has one output and two or more inputs. The output will stand at its defined logic 0 state if one or more of the inputs stand at their defined logic 1 states. *See Figure N.3.*

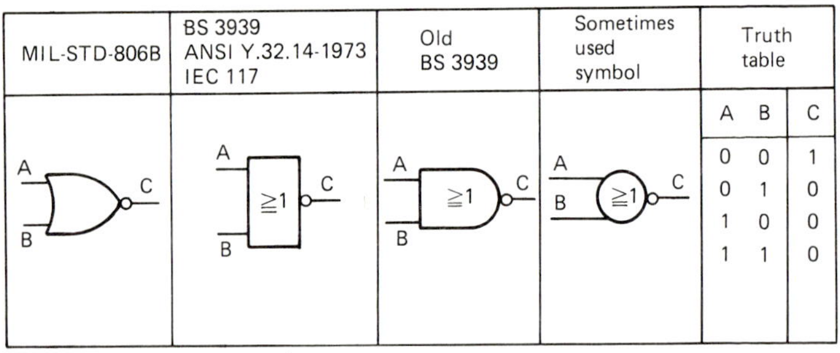

| MIL-STD-806B | BS 3939 ANSI Y.32.14-1973 IEC 117 | Old BS 3939 | Sometimes used symbol | Truth table | | |
|---|---|---|---|---|---|---|
| | | | | A | B | C |
| A ─ C, B (≥1 NOR symbol) | A ≥1 C, B | A ≥1 C, B | A ≥1 C, B | 0 | 0 | 1 |
| | | | | 0 | 1 | 0 |
| | | | | 1 | 0 | 0 |
| | | | | 1 | 1 | 0 |

*Figure N.3* The NOR gate. In practice the MIL-STD-806B symbol is the one most likely to be met, although the other logic symbols may be encountered

**NOS** Network Operating System. A generic term employed to describe a distributed software for the control of the operation of a computer network.

**NOT** A logical operator having the property that if A is a statement, then the NOT of A is true if A is false and it is false if A is true.

**NOVRAM** NOn-Volatile Random-Access Memory. A particular combination of random-access memory (RAM) and read only memory (ROM) which produces a device with RAM attributes that is unaffected by the loss of power, in that it does lose its contents. It is therefore termed non-volatile. (*See* NVM, non-volatile memory).

**n-p** N-doped P-doped junction. A region in a semiconductor where the n-doped and p-doped areas meet. It has a high resistance in one direction and a low resistance in the other.

**Np** Neper. A dimensionless unit used in telecommunications. It is the natural logarithm of the square root of the power ratio and is named after the inventor, John Napier.

**NPA** Numbering Plan Area. Part of a national or integrated telephone numbering plan, it is an area that consists of further numbered subdivisions that are identified by a particular number-plan trunk code. In the UK, London has been given the digit 1.

**npin** N-doped P-doped Intrinsic N-doped transistor. A transistor that is constructed with an intrinsic (undoped) layer of material between its base and collector to extend the high frequency range.

**npip** N-doped P-doped Intrinsic P-doped transistor. A transistor that has an intrinsic (undoped) layer between two p-regions (p-type).

**NRCd** National Reprographic Centre for Documentation. A UK organization devoted to the study of reprographics, publishing information and evaluating equipment.

**NRMM** National Register of Microfilm Masters. A register of original negatives of microfilms or microfiches housed in the US Library of Congress.

**NRZ** Non-Return to Zero. (a) A signal presented in the digital (D,4) form having pulses which are full-symbol-length (*see* diagram) so that there is no alteration of signal level when a symbol is repeated, i.e. there is no return to zero level, as shown in *Figure N.4*. (b) A method of writing information on a magnetic surface where the current passing through the write head winding does not return to zero after the write pulse. The reference condition is represented by the change in state of magnetization between the 1 and 0 condition.

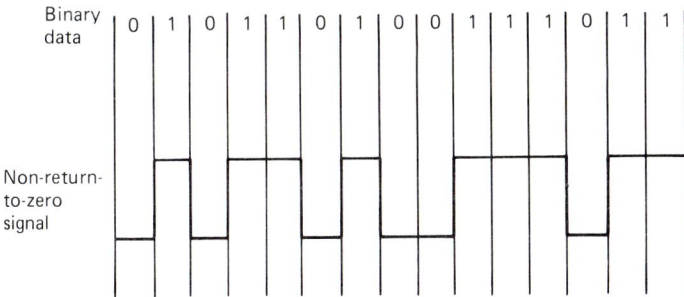

*Figure N.4*  A non-return-to-zero (NRZ) signal

**NRZI** Non-Return to Zero Inverted. A method of writing information on a magnetic surface in which the state of magnetization representing either 0 or 1 changes only when a 1 is transmitted.

147

**NRZ(M)** Non-Return to Zero (Mark). A method of writing information on a magnetic surface in which a change in the condition of magnetization represents a 1 and no change represents a 0.

**ns** Nanosecond. *See* nsec.

**nsec** Nanosecond. One thousandth of a millionth of a second (s) that is 0.000 000 001 or $10^{-9}$ seconds. It is synonymous with millimicrosecond.

**NSI** National Supervisory Inspectorate. A controlling authority, with inspectors at local level, which is the supervisory body for the approval and quality assessment of connectors and other components to British Standard 9000 series specifications (*see* BS 9000).

**NSW** National Software Works. A co-operative venture amongst some associations in the USA, to design a distributed operating system (OS) for a heterogeneous computer network.

**NTC** Negative Temperature Coefficient. The attribute of a semiconductor material which has a resistance that decreases with an increase in temperature. NTC semiconductor materials are normally employed in thermally sensitive resistors (THERMISTOR).

**NTIA** National Telecommunications and Information Agency. An agency of the Department of Commerce in the USA.

**NTIS** National Technical Information Service. Covering technical and scientific reports from US Government agencies, this service generates a multi-disciplinary store of information on computer files.

**NTSC** National Television System Committee. An American body that specified and pioneered the NTSC compatible colour television system in 1953, the system being used in the USA and in a number of other countries. The Phase Alternation Line (PAL) system in the UK and other countries is an improved development of the NTSC system. The body also determines standards on behalf of the USA Standards Institute (USASI).

**NTU** Network Terminating Unit. Part of telecommunication network equipment that connects directly to data terminal equipment. It operates between local transmission lines and a telephone subscriber's interface.

**n-type** (1) An extrinsic semiconductor containing a higher density of conduction electrons than of mobile holes (Valence band electrons), which means the electrons are the majority carriers. A semiconductor region is doped to provide the excess of electrons (*see Figure N.5*). Compare p-type semiconductor. (2) An early type of microwave connector. It is screw-coupled and has an air interface, the centre conductor being supported

148

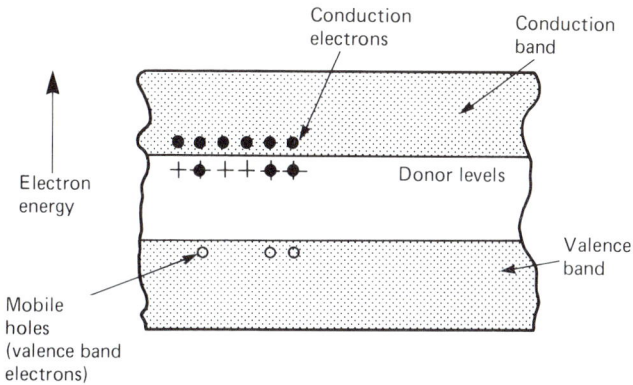

*Figure N.5* Energy bands of an n-type semiconductor

by a polytetrafluoroethylene (PTFE) or polystyrene bead.

**nu** The nu value or abbé constant is a mathematical expression applied to optical systems for determining the correction required to counteract image imperfections in the system.

**NU** Number Unobtainable. A tone used in telephone communications indicating to a caller, by use of an audible signal, that a call cannot be completed. There are four possible reasons: the line required is either faulty or out of commission; a non-existent or invalid number has been dialled; equipment has been misoperated or has malfunctioned; access to the required number is banned. The tone is also relayed if a caller does not send an effective signal during a certain period of time.

**NUA** Network User Address. A set of packet assembler-disassembler (PAD,2) parameters that identifies the computer with which a local area network (LAN) wide area network (WAN) user wishes to communicate.

**NUI** Network User Identity. A set of packet assembler/disassembler (PAD.2) parameters that provide a local area network (LAN) or wide area network (WAN) user with a recognizable identity.

**NUL** NULl (a) The absence of information, as opposed to a zero or blank which indicate the presence of no information. (b) Zero. (c) The deflection from a centre or end position. (d) A computer instruction to do nothing, except proceed to the next sequential operation. Null characters can be inserted into, or removed from, a sequence of characters without the meaning of the sequence being affected although the control of equipment or the format may be altered.

**NVM** Non-Volatile Memory. A permanent memory storage medium that is able to retain information in the absence of

149

power and allows the data to be available when power is restored. A magnetic memory is an example of a non-volatile memory.

# O

**O** (1) Oxygen. A chemically active, gaseous element forming one fifth of the Earth's atmosphere. It is odourless, has an atomic number 8, atomic weight 15.9994, melting point $-218.8°C$ and a boiling point $-182.970°C$.

(2) Output. (a) Results issuing from a computer or device which could be answers to statistical or mathematical problems. (b) Signals produced by a device that are used to drive peripheral devices. (c) Information transferred from the internal storage of a computer to secondary or external storage. (d) A channel for expressing a state on a device or logic element.

(3) O network. A network consisting of four impedances connected to form a closed loop or square (*see Figure O.1*). It has an input applied across one element and the output (O,2) taken across the opposite element.

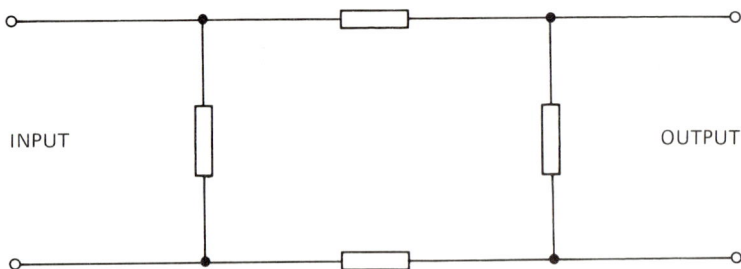

*Figure O.1* An O network

**OC** Optical Cavity. The active area or cavity in a laser, formed by two heterojunctions (HJ). *See* LOC, large optical cavity.

**OCB** Oil Circuit-Breaker. A high-voltage (HV) circuit-breaker that has an arc through oil, which raises the temperature of the oil causing bubbles of gas (mainly hydrogen) to be formed. When the current is returned to zero the turbulence caused by the gas and the cooling effect of the surrounding oil disperses and de-ionizes the path of the arc and restores its dielectric strength.

**OCLC** Ohio College Library Centre. An on-line computer network, with more than 3000 remote terminals, for cataloguing library information. It is available mainly in North America.

150

**OCR** (1) Optical Character Reader. A device used to produce coded signals, in digital (D,4) form, from printed characters, numbers or symbols on paper, that are suitable for a given computer. It usually requires source material in a specially designed typeface for maximum efficiency.

(2) Optical Character Recognition. Printed or written characters recognized by machines using inputs from optical, photoelectric transducers or other photosensitive devices.

**OCR WAND** Optical Character Reader WAND. A hand-held optical character reader (OCR,1). An example is a bar code scanner, which is a unit that can read documents encoded in a special bar code, at speeds of hundreds of characters per second (CPS).

**OCT** OCTal. Pertaining to eight, but usually used to describe a numbering system with the base eight. Numbers in the system express values with each digit position representing successive powers of eight. For instance, decimal 85 would be written as 125, representing 1 times 64, 2 times 16 and $5 \times 1$. If used as a shorthand method of writing binary digit (BIT) strings, the bit string 001,010,101 can be written as octal 85, whereas in straight binary it is 1010101.

**ODB** Output to Display Buffer. Peripheral to a computer, this data storage area holds information whilst it is being transmitted from an output (O,2) to a display (D,6) device.

**Oe** OErsted. In the now obsolete centimetre, gram, second (CGS) system of units, this is the unit of magnetic field. One oersted is equal to 79.58 amperes (A) per minute.

**OE** Output Enable. An output (O,2) control signal that provides a completely separate means of controlling the output buffer of a memory device, so eliminating the need for bus connections.

**OEM** Original Equipment Manufacturer. A manufacturer who uses computers or components as a section or part of the items that he directly sells, as opposed to end users who buy the products for their own use.

**OF** Optical Fibre. A dielectric waveguide for electromagnetic energy at optical wavelengths. The fibres provide a path for a single beam of light or in multiples, such as the transportation of a complete image. The fibres are provided as either a single fibre or a cable bundle. They may be bent or curved to meet specific needs. *See* FO, (fibre optics).

**OF LIN PS** Open Frame LINear Power Supply. A low cost, very simple construction power supply unit (PSU) that employs linear techniques. Normally mounted on a single open chassis without a cover, this type of unit has a power rating of up to 150 watts (W,2).

**OFPS** Open Frame Power Supply. A power supply unit that is normally mounted on a single chassis without a cover. It is a very cost-effective design intended for original equipment manufacturers (OEM) to build into their own equipment.

**OF SMPS** Open Frame Switch Mode Power Supply. A switch mode power supply (SMPS) that is normally mounted on a single open chassis without a cover. It is a very cost-effective design intended for original equipment manufacturers (OEM) to build into their own equipment. *See Figure O.2.*

*Figure O.2* An open-frame switch mode power supply (OF SMPS) (*courtesy* Coutant Electronics Ltd)

**OFTF** Optical Fibre Transfer Function. A constant for a particular optical fibre (OF). When a light wave enters an optical fibre the fibre brings about a transformation upon it. If the input composition and the transfer function are known, then the output can be determined.

**OIS** Office Information System. A general term that can be applied to any of the electronic systems employed in a variety of office functions. The processes concerned can include information retrieval, telecommunications or word processing (WP).

**OLRT** On-Line Real-Time operation. A method of operation whereby results from a computer are obtained during the progress of an event. Input data are passed directly from a measuring device to the computer.

**OLTS** On-Line Test System. The running of a whole system against test data with a verified solution, or the complete simulation of an actual running system to test out the adequacy of the system. The system runs at speeds compatible with a computer and is connected directly to the main processing unit (MPU).

**OMNIX** Onyx Microcomputer uNIX. A version of the general-purpose multi-user interactive operating system (OS) (*see* UNIX) available on Onyx microcomputers.

**OMR** (1) Optical-Mark Recognition. A technique in which a device scans and recognizes graphite marks on a document and converts the information into digital (D,4) signals for input to a computer.

(2) Optical-Mark Reader. A device that recognizes graphite marks, which are normally produced by a pencil on card or paper, and converts the information into digital (D,4) signals for input to a computer.

**ONI** Operator Number Identification. Equipment at a local dial central office that permits an operator to break-in for sufficient time to identify and acquire the calling telephone number. This is done so that the number may then be dialled into centralized automatic message accounting equipment.

**ON-TAP** ON-line Training And Practice. A specific set of

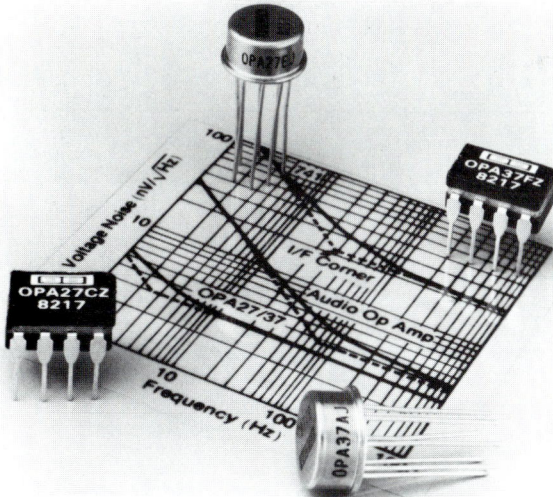

*Figure O.3* Examples of precision operational amplifers (OP AMP) (*courtesy* Burr-Brown International Ltd)

information held on computer files containing data from a number of other stores of information held on computer file. It is designed to help train personnel in the uses of direct searching of computer files.

**OP AMP** OPerational AMPlifier. Normally an integrated amplifier (AMP) of high gain and wide bandwidth that can be used to perform mathematical operations. *See Figure O.3.*

**OP REGISTER** OPeration REGISTER. A memory storage location used to hold the operation code of computer instructions.

**OPM** Operations Per Minute. The number of defined actions executed in one minute.

**OPR** Optical Pattern Recognition. A technique in which an optical scanning device is used to convert scanned information into digital (D,4) signals. The method is normally used for the computer recognition of patterns by comparison to a known reference, for classification, identification or sorting.

**OR** (1) Operational (or Operations) Research. Scientific methods applied to the solving of problems that concern the allocation of resources. They make use of mathematical models and analytical methods to provide logical basis for solving operational problems as well as for making sound predictions and decisions that have been designed for and are used by management.

(2) OR gate. This logic element has one output (O,2) and two or more inputs. Its output will stand at its defined logic 1 state if one or more of its inputs stand at their defined logic 1 state. *See Figure O.4.*

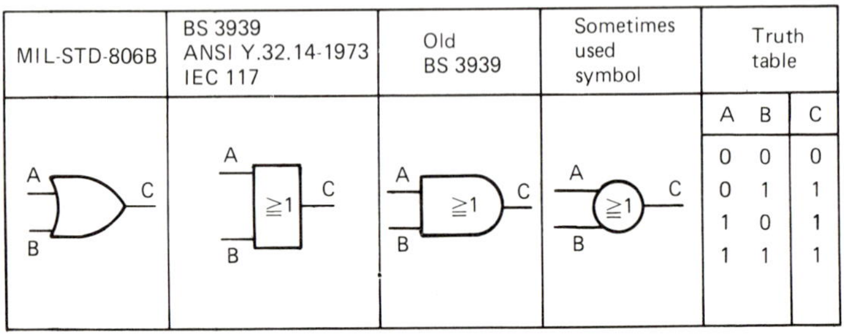

| MIL-STD-806B | BS 3939 ANSI Y.32.14-1973 IEC 117 | Old BS 3939 | Sometimes used symbol | Truth table | | |
|---|---|---|---|---|---|---|
| | | | | A | B | C |
| | | | | 0 | 0 | 0 |
| | | | | 0 | 1 | 1 |
| | | | | 1 | 0 | 1 |
| | | | | 1 | 1 | 1 |

*Figure O.4* The OR gate. In practice the MIL-STD-806B symbol is the one most likely to be met, although the other logic symbols may be encountered

**ORACLE** The name given by the Independent Broadcasting Authority (IBA) for its broadcast information service. It consists of the transmission of information in a coded form within the standard television signal, thus providing text display (D,6) or pictorial matter on the screen of receivers that have the necessary circuitry for reception (teletext systems).

**ORBIT**  On-line Real-time Branch Information. A computer language developed by International Business Machines (IBM) to search stores of information held on computer files.

**ORGALIME**  ORGAnization for LIaison between electrical and Mechanical Engineering industries. An association that represents the national trade organizations for electrical and mechanical engineering, of the European Economic Community (EEC) and European Free Trade Association (EFTA) countries.

**ORSA**  Operations Research Society of America. A body that is interested in extending the computer far beyond automatic and routine operations, into decision making and other complex processes of management.

**OS**  Operating System. (a) The organized collection of procedures and techniques that are required for the operation of a computer. (b) That part of computer software necessary to simplify administrative procedures for input/output (I/O), data-conversion and other routines.

**OSCAR**  Orbital Satellite-Carrying Amateur Radio. The name of a number of satellites launched by The Radio Amateur Satellite Corporation (AMSAT).

**OSD**  Optical Scanning Device. A technique in which a device scans graphics or text and converts the scanned information into digital (D,4) signals for computer processing.

**OSDM**  Optical Space-Division Multiplexing. Bundles of independent fibres that provide optical paths for independent channels (CHNL(d)). In practice this is a single cable with overall sheathing and strengthening made from a collection of fibres. *See* SDM, space-division multiplexing.

**OSI**  Open System Interconnection. A term used generally to describe the technique employed for the construction and connection of a computer network, in which the types of interconnected computer systems employed are different. Proposed specifications of such a system have been drafted by the International Standards Organisation (ISO).

**OTD**  Optical Time Domain. A distance, measured in time, from a light source to a reflective surface and back again. The concept is used for checking fibre-optic (FO) cables with an OTD reflectometer. The reflectometer launches a light pulse into a fibre-optic cable and measures the time taken for its reflection to return, thus indicating continuity or discontinuity caused by cracks, fractures or breaks.

**OTP** Office of Telecommunications Policy. An executive office of the US President that has no statutory powers of regulation but develops and recommends public policy in the area of telecommunications.

**OTS** Orbital Test Satellite. A programme started in 1977 by the European Space Agency for experimental communication satellites (COMSAT).

**OTS II** Orbital Test Satellite II. A geostationary satellite launched in 1978 as a test station for the equipment, techniques and systems intended for further use by the European fixed-service satellite system (ECS) of the 1980s. OTS II has the capacity to deal with 6000 telephone calls and two television (TV) channels (CHNL(g)) simultaneously. OTS I was never brought into commission due to failures during its launch.

**OV** OVerflow. An overcapacity, for instance the generation in an arithmetic operation of a quantity beyond the capacity of the memory store or location to receive it. Fields that are longer than field locations will overflow, e.g. a 10-digit product will overflow an 8-digit accumulator (ACC).

**OVD** Optical Video Disk. A disk having on its surface a spiral or concentric circles of recorded digital (D,4) data at a high packing density (D,3). The information is recorded in spots using a laser beam and is read by reflective means using a low intensity laser beam. Using a suitable machine the disk can be used to provide television (TV) programming playback.

**OVP** Over-Voltage Protection. The action of defending a circuit (CCT) from voltages (V,2) larger than predetermined requirements. *See* OVPU, over voltage protection unit.

**OVPO** Outside Vapour-Phase Oxidation. A chemical vapour-phase oxidation (CVPO) process that is used for the production of optical fibres (OF). Dopants are burnt with oxygen (O,1) and gas to form an oxide stream that is deposited onto the outside surface of a rotating glass rod.

**OVPU** Over-Voltage Protection Unit. A circuit or device that comes into operation when the supply voltage (V,1) in a particular circuit (CCT) rises above a predetermined value. Such units are used to protect computers and delicate equipment from short-duration transients (spikes) in supply voltages.

**OWF** Optimum Working Frequency In high frequency (HF) transmissions, an OWF is a frequency that is lower than the maximum useable frequency (MUF) by an amount sufficient to allow for expected variations in the propagation conditions that may occur during the working period of operation.

**p**  (1) Pico. A decimal prefix commonly used in association with a base unit in the SI system of units indicating a multiplication factor of $10^{-12}$ of that unit. For example, one picofarad (pf) is equal to 0.000 000 000 001 farad.

(2) P-channel. A metal oxide semiconductor (MOS) circuit in which the conducting path is formed as a p-doped (p-type) semiconductor. This type of channel (CHNL(h)) can also be found in junction field-effect transistors (JUGFETs).

**P**  (1) Peta. A decimal prefix used in association with a base unit in the SI system of units indicating a multiplication factor of $10^{15}$ of that unit. For example, one petatonne, PT, is equal to 1 000 000 000 000 000 tonnes.

(2) P band. A band of microwave frequencies that range from 0.225 to 0.390 gigahertz.

(3) Power. The rate at which work is done or energy is expended; the unit is the watt (W). In a direct current (DC) circuit or device, $W = VI$, where V is the potential difference (PD) in volts (V) and I the current (I) in amperes (A).

**Pa**  PAscal. The SI unit of pressure. It can be defined as the pressure that results when one newton (N) acts uniformly over an area of one square metre (m).

**PA**  (1) Public Address (system). Usually consisting of one or more microphones (MIKE), power amplifiers (AMP) and several loudspeakers, PA systems are designed to reproduce sounds, most often speech or music, so they become audible to a large gathering of people. They are used for airport announcements, in railway stations, open-air meetings or large halls.

(2) Paper Advance. Usually a key or button provided on a printer that moves the paper or printout forward independently of the action of the printing head or impressing device.

**PABX**  Private Automatic Branch eXchange. A privately owned automatic branch telephone exchange that provides access to and from the public telephone network without operator intervention.

**PAC**  Program Address Counter. A sequential counter that is able to keep track of the location of the next instruction to be executed from a program memory.

**PACC**  Product Administration and Contract Control. A particular method of control employed by a data management system in a business environment.

**PACVD**  Plasma-Activated Chemical Vapour Deposition. A method for producing graded index (GI) optical fibre (OF) using a form of the chemical vapour deposition (CVD) process. Thin layers of different indices are deposited on the inner wall of a glass tube, a chemical vapour flows through the tube and the formation of oxides is stimulated by means of plasma generated by a microwave ($\mu$W) resonant cavity.

**PAD** (1) Packet Assembly/Disassembly. Provided in a packet-switching exchange, a PAD facilitates communication between two terminals of a different nature. It has two basic functions, firstly to assemble received data into packets for onward transmission and secondly to disassemble these packets and transmit to various other terminals.

(2) Packet Assembler/Disassembler. A unit that provides a packet assembly/disassembly (PAD,1) facility.

**PAL** Phase Alternation Line. A system used in the UK, most of Europe, and elsewhere, for 625-line colour television (TV) transmission. Signals are much less susceptible to transmission problems than in other systems, as the technique that is used inverts the phase of one of the two chrominance signal components every alternate line.

**PAL D** Phase Alternation Line Delay. A method of decoding signals used in the Phase Alternation Line (PAL) colour television (TV) system, that reduces colour errors.

**PALLNIC** PALLadium-NICkel. A palladium/nickel alloy of between 65% and 85% palladium content, with a colour and finish similar to that of bright stainless steel. It was developed to fulfil needs within the electronics industry to replace gold in connectors and printed circuit board (PCB) edge connectors.

**PAL S** Phase Alternation Line Simple. A process of decoding signals used in the Phase Alternation Line (PAL) colour television (TV) system, that requires the observer by sight to optically integrate any switched, line-by-line colour errors.

**PAM** Pulse Amplitude Modulation. A form of modulation used in communications, where the amplitude of the pulse carrier is varied in accordance with successive samples of the modulating signal.

**PANDA** Prestel Advanced Network Design Architecture. A particular design of construction for interconnecting Prestel devices.

**PAR** Precision Approach Radar. An aid to air traffic control, used at an airport to provide accurate information concerning the location of incoming aircraft in the local vicinity. It is usually used in association with an airport surveillance radar (ASR).

**PARD** Periodic And Random Deviation. A term recommended by the International Electrotechnical Commission (IEC) as the specification for ripple and noise.

**PASCAL** (1) A computer language designed originally to enable the teaching of programming as a systematic discipline and to allow those being taught to do systems programming. It is based on an algorithmic language (ALGOL). First developed at the Zurich Technical Institute in the late 1960s it emphasizes aspects of structured programming.

(2) Program Applique à la Sélection et la Compilation Automatique de la Literature. Compiled by the Centre National de la Recherche Scientifique in France, this is a very large store of information held on computer files. It covers a wide range of different subjects.

**PATCH** A colloquial expression (a) normally employed to describe a non-standard or unusual correction to a computer program that corrects a bug or unexpected function, (b) employed to describe an extract connection or link that was not intended at the design stage but enables a circuit (CCT) or a printed circuit board (PCB) to function.

**PATRICIA** Practical Algorithm To Receive Information Coded In Alphanumeric. A set of simple procedures or rules for receiving alphabetical and numerical information.

**PATSY** Programmer's Automatic Test SYstem. A procedure that assists people who prepare programs for a computer, in debugging routines.

**PAU** Pattern Articulation Unit. A microprocessor (MP) that is capable of converting graphics images into a stream of data or information. It can also reverse the process, i.e. reconstruct graphics images from converted information.

**PAX** Private Automatic eXchange. A dialled telephone exchange that provides a private service within an organization, but does not allow calls to be transmitted to or from the public telephone network.

**PB** Peripheral Buffer. An input/output (I/O) buffer situated within the memory store of a peripheral device. It permits data-word transfers to and from the computer memory without the need of main program attention.

**P band** P band microwave ($\mu$W) frequencies occur between 0.225 and 0.390 gigahertz (GHz) and form part of a larger continuous range of frequencies covering 0.225–100 gigahertz which are usually subdivided into bands designated by letters. The bands are commonly used subdivisions but, however, are not internationally agreed. P band frequencies have wavelengths of between 133.3 and 76.9 cm.

**PBJ** Paper-Braided Jute. An insulation material for cables.

**PBX** Private Branch eXchange. A private telephone exchange located on a subscriber's premises that has access to the public telephone network. The term can be applied to automatic or manual exchanges, although in the UK an automatic exchange is referred to as a PABX and a manual exchange is referred to as a PMBX.

**PC**   (1) Printed Circuit. A circuit (CCT) in which all the interconnecting wires have been replaced by conductive strips, normally copper. These strips have to be plated, printed or etched onto an insulating baseboard. *See* PCB, printed circuit board.

(2) Personal Computer. A small low-cost computer (relative to standard and mini systems) available as ready-built units or in kit form. Applications include professional business, small business, hobbyists and home, the individual software being orientated towards single user applications. The two main categories are (a) those that have their own cathode-ray tube (CRT) display (D,6) and cassette or floppy disk (FD,1) facilities for program storage and exchange; and (b) those that rely on a domestic television (TV,(c)) set as an output (O,2) display and an audio cassette for program storage and exchange. Both are able to accommodate peripheral devices such as standard or purpose designed printers.

(3) Portable Computer. A hand-held micro or personal computer (PC,2). Typically these devices can accommodate add-on printers or cassette units.

(4) Program Counter. A processor register that is automatically advanced, usually by 2, every time the processor retrieves an instruction. This information is then used in connection with operating modes.

**P-C**   Processor-Controller. With reference to computer systems, a processor-controller can be used for direct control, data analysis, supervisory control or editing.

**PCB**   Printed Circuit Board. A circuit (CCT) or part of a circuit formed on an insulating board or card, in which the conducting interconnecting paths have been printed or etched. PCBs may be single-sided, double-sided or multilayer for high-density applications.

**PCB PS**   Printed Circuit Board Power Supply. A power supply unit (PSU) mounted on a single printed circuit board (PCB) or card. *See Figure P.1.*

**PCI**   (1) Panel Call Indicator. Relays or switches at a terminating office that register the detection of a direct current (DC) pulsing system. In this system, information digits (DGTs) are each transmitted as a series of four marginal, polarized pulses.

(2) Program Check Interruption. Caused by unusual conditions

160

experienced in a computer program, this interruption is usually due to an incorrect operand.

(3) Process Control Interface. Manufactured in the form of a circuit on a board, this device can provide computers with an interface to a variety of electromechanical devices such as relays, switches, motors, contacters, thermostats and various other devices.

**PCM** (1) Pulse Code Modulation. A form of modulation used in communications where the modulating signal is sampled and the sample quantized and coded so that each element of the information consists of one or more binary digits (BITs). In a communication system this type of signal modulation allows noise-independent transmission, as the signal is made up of a rapid sequence of pulses which have the same amplitude, arranged in binary groups. These correspond to numerical groups which represent the amplitude values needed to recreate the slope of a signal or the pulses can represent data bits.

(2) Punched (Punch) Card Machine. Widely used items of equipment that punch, read, analyse or deal in any way with, punched cards. These include sorters, collators and tabulators.

(3) Plug Compatible Manufacturer. A manufacturer of peripheral equipment that produces items fully compatible with a particular computer, the computer being made by another manufacturer. The devices simply plug into the computer.

**PCMI** Photo-Chromic Micro Image. An ultrafiche having extremely small images such that over 3000 pages can be held on an A6 card. It is similar to a microfiche, but requires much higher standards of control during production.

**PCP** (1) Plug Compatible Peripherals. Peripherals to a computer, which can include main memory, that simply plug in. The PCPs are made by a different manufacturer than that of the computer.

(2) Primary Control Program. A particular sequence of instructions that prescribe the series of steps to be taken by a computer system, the system having been designed to perform one task or job at a time.

**PCS** (1) Personal Computing System. A relatively low cost portable computer that has software designed for single-user applications, personally controllable and easily used. There are several classifications of systems, which include home, hobbyist, professional and small business.

(2) Plastics-Clad Silicon. Optical fibre (OF) with a core of pure silica glass covered by an overall plastics cladding. Its loss and dispersion are normally higher than other types of fibre-optic (FO) material.

(3) Print Contrast Signal. A value of a measured difference between the contrast of a character that has been printed and the contrast of the paper on which that character appears.

(4) Punched-Card System. A system for encoding information onto cards in the form of holes. The pattern of positions punched for any given character is determined by the particular code employed by the card punch.

**PD** (1) Potential Difference. The SI unit of potential difference is the volt (V). It can be defined as the line integral of the field strength along a freely chosen path between two points in an electric field that is produced in the absence of electromagnetic induction.

(2) Photo Detector. A light-detecting device that can be used for communications by extracting information from an optical carrier.

**PDA** Problem Determination Aid. A device or routine that assists the procedure which is to be adhered to in order to determine the cause of an error.

**PDD** Post Dialling Delay. A time interval that occurs between the finish of dialling on a telephone receiver and the reception of a return signal, which could be either a ringing tone or a busy tone (BT,2).

**PDL** Print Definition Language. Used by a computer to define printer characteristics including lines, columns, page and overflow print conditions.

**PDM** Pulse Duration Modulation. A particular form of pulse modulation (PM) in which the duration of each pulse in a pulse train is varied according to the instantaneous value of the modulating wave. The modulating signal can vary when the leading edge, the trailing edge or both edges of each pulse occurs.

**PDP** Programmed Data Processor. A wide range of computer system, from large mainframes to minicomputers, manufactured by the Digital Equipment Corporation.

**PDS** Partitioned Data Set. A computer file that is divided into subfiles which are able to be referenced as a group or individually.

**PDU** Pilot Display Unit. A display (D,6) unit mounted in a cockpit on which information is projected in the pilot's line of sight and superimposed over his view of the real world. *See* HUD, head-up display.

**PE** Phase Encoding. A particular technique employed for encoding information onto a magnetic tape (MAGTAPE).

**PEC** (1) Photo-Electric Cell. A light-sensitive device comprising two electrodes in a glass bulb that can be evacuated or filled with an inert gas. The unit provides an electrical output (O,2) directly dependent upon the amount of light falling on its input electrode.

The term more commonly used is 'photocells' and these are often made from semiconductor materials. There are several different types of photocell which can be either photo conductive, or photovoltaic, or photo-emissive.

(2) Panel Electronic Circuit. A printed wiring panel that has components mounted upon it.

**PEL** Picture ELement. The smallest detail in a picture area, comprising a television (TV) facsimile (FAX) or other transmitted picture, that can be resolved by the scanning system without loss of amplitude; also known as 'pixel'. *See* PIXEL.

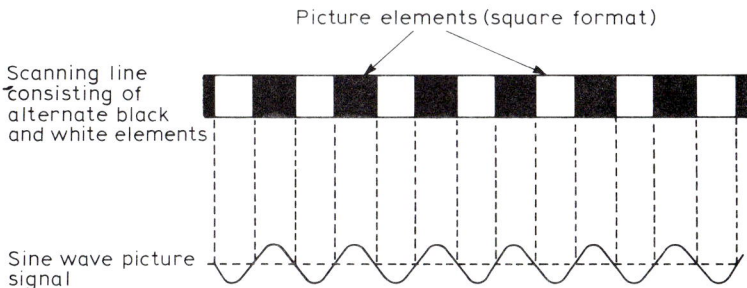

*Figure P.1* A picture signal corresponding to a row of alternate black and white picture elements (PEL)

**PEMS** Porcelain Enamelled Metal Substrates. Replacements for printed circuit boards (PCBs) and thick film circuits on ceramic. Circuit boards using PEM substrates act as their own heat sink/electrical ground plane; are not affected by severe conditions of humidity, temperature and thermal or physical shock; and may be formed into a wide variety of shapes to meet specific applications.

**PEP** Peak Envelope Power. The mean power supplied to an antenna during one complete radio frequency (RF) cycle at the peak of the modulation envelope.

**PEPE** Parallel Elements of Processing Ensemble. An early method of executing a number of processes in a computer, in parallel.

**PERIPHERAL** PERIPHERAL device or equipment. An auxiliary unit or piece of equipment of a computer system. The devices normally perform input/output (I/O) operations, such as disk drives or paper tape readers (PTR).

**PERMALLOY** PERMeable ALLOY. An alloy with a high

163

magnetic permeability (at low values of magnetic flux density), low coercivity, and therefore low hysteresis loss. It contains 78.5% nickel and 21.5% iron, although varieties are produced by adding copper, cobalt, manganese and other metals.

**PERT** Program Evaluation and Review Technique. A computer program that provides a method of analysis and reporting for overall project administrators. Areas that impose the greatest time restriction for completion of a project as well as those with excess time for completion can be highlighted. Similar to critical path method (CPM).

**PET** Personal Electronic Transaction computer. A popular personal home computer, based on the 6502 8-binary digit (BIT) microprocesor (MP) manufactured by Commodore Business Machines.

**pf** PicoFarad. One millionth of a millionth of a farad (F), that is 0.000 000 000 001 or $10^{-12}$ farads.

**PF** (1) Pulse Frequency. The number of complete pulses measured during a particular time interval.

(2) Page Footing. A record of information or data situated and printed at the bottom or foot of a page(s) of a document.

(3) Packing Fraction (or Packing Factor). A ratio of the active core total cross-sectional area of a bundle of optical fibres (OF) or cable cores, to the total cross-sectional area of the bundle. The packing fraction can be expressed as $PF = N(A/D)^2$ where N is the number of cores, A is the diameter of each core and D is the diameter of the whole bundle or assembly.

(4) Power Factor. A comparison of average power dissipation against the apparent power in an alternating current (AC) network or part of that network. It is expressed as a ratio.

**PFM** Pulse Frequency Modulation. A form of pulse modulation (PM) in which the pulse repetition rate is varied.

**PFR** Power Fail Recovery (or Power Fail Restart). A system containing automatic switching and charging circuits, designed to protect computer systems by detecting a drop in input voltage (V), which signals an imminent power failure. This provides several tens of milliseconds (ms) during which all the memory locations can be maintained by a battery backed-up memory, or preserved in a non-volatile memory and then automatically restarted when the power is restored.

**PGA** Pin Grid Array. A mounting technique employing a high density package for very-large-scale integration (VLSI) integrated circuits (IC).

**PGS** Program Generation System. Allowing the user to select a particular area of memory and output (O,2) it in a particular format, this system will then load programs into memory. It can load self-produced programs and also others produced elsewhere.

**pH** Percentage Hydrogen. A measure of acidity, expressed as the common logarithm of the reciprocal of concentration.. It indicates the concentration of hydrogen ions in a solution.

**PH** Page Heading. A record of information or data that is situated and printed at the top or head of a page or pages in a document.

**pi** PI-network (or $\pi$-network). A particular type of network or circuit (CCT) comprizing two equal shunt arms with a series arm between them.

**PI** (1) Programmed Interrupt. Often used as a fast method of transferring control to an error routine, a PI is caused deliberately by a computer program (as opposed to one caused by an external event).

(2) Paper Insulated. Generally refers to power cables that use insulating paper having a density of between 0.7 and 0.9 g/cm$^2$ when dry. The paper is impregnated with insulating oil to improve its dielectric strength.

**PIA** Peripheral Interface Adaptor. A device that will permit parallel orientated peripherals to be connected to various microprocessors (MP) usually without any additional circuits.

**PIC** Priority Interrupt Control. A special chip (CHIP) in a microcomputer that manages several, normally eight, external attention signals sent from an input/output (I/O) device or chip to the microprocessor unit (MPU) to obtain service.

**PICTEL** PICture TELephone. A name used by the US General Telephone and Electronics Company to describe its experimental video telephone that permits the user to see as well as talk with another party.

**PID** Peripheral Interface Device. Interface printed circuits, compatible with a set of standard peripherals, that can all be plugged into a chassis assembly. It may sometimes be a separate unit.

**PIL** Precision-In-Line. A colour picture tube employing a shadow mask near the screen in which there are vertical slits. The screen has vertical stripes of red, blue or green phosphor dots. Three electron guns or one that fires three beams are used and the mask ensures that any one strip is hidden from two of the beams.

**PILOT** An interactive computer-aided instruction (CAI) computer programming language developed by John Starkweather at the University of California.

**PIM** Pulse Interval Modulation. A form of pulse modulation (PM) in which constant amplitude pulses of constant width have their spacing varied.

**pin** or **p-i-n** Positive-Intrinsic-Negative. A junction semiconductor diode (D,5) with an area of semiconductor material nearing intrinsic conductivity (i-type) between the positive p-doped semiconductor material (p-type) and the negative n-doped semiconductor material (n-type). At low frequencies the diode acts similarly to a p-n semiconductor junction (p-n), but at high frequencies it exhibits variable resistance and is suitable for impact ionization avalanche transit time (IMPATT) operation. P-i-n diodes are used as photodetectors (PD).

**PIN** *See* pin.

**PIO** (1) Parallel Input/Output. An interface that permits a computer to input and output (O,2) data bits simultaneously to and from external parallel devices, such as a visual display unit (VDU) or a keyboard (KB). It is a circuit (CCT) in a computer system or a separate unit that is bus compatible with various microcomputers and peripheral devices. Most are programmable to operate in static, clocked (CLK) or discrete modes.
(2) Programmable Input/Output chip. An input/output (I/O) chip (CHIP), in most cases an 8-binary digit (BIT) interface chip that multiplexes (MUX) a connection to a data bus, which is capable of transferring data to and from the central processor unit (CPU) and peripheral devices, into two or more ports.

**PIP** (1) Peripheral Interchange Program. A series of computer programs that transfer information from one device to another.
(2) A significant intensification or deflection of the spot on a cathode ray tube (CRT), that is providing a display (D,6) for identification or calibration. Normally it is applied to the peaked pattern of a cursor signal or device.

**PIRS** Personal Information Retrieval System. A software package used together with a mini or microcomputer to order and index information for storage on a magnetic medium.

**PIT** Programmable Interval Timer. A chip (CHIP) that has a separate timing clock (CLK) and a number of special memory storage locations. These are used independently of the microprocessor unit (MPU) to count time for real-time applications. They can store the time-elapsed or generated signals, at the completion of a time period.

**PIU** Programmable Interface Unit. A programmable large scale integration (LSI) device that is used to eliminate the hardware logic often needed for communication interfaces, between processors and various other system components.

**PIV** Peak Inverse Voltage. The maximum instantaneous voltage (V) across the input terminals of a diode (D,5) in a rectifying circuit, when the diode is non-conductive.

**PIXEL** PICTure ELement. A small rectangular division of a video screen. More, and smaller, rectangles are necessary to achieve higher resolution. Therefore a sharper pictures requires more computer memory to store the pixels. *See* PEL.

**PL/1** Programming Language ONE. Developed by International Business Machines (IBM) in the mid-1960s, this is a high level computer language employed in a varied range of scientific and commercial applications. A version now exists for microcomputers.

**PLA** Programmable Logic Array. An orderly arrangement of logical AND (*see* AND) functions and logical OR (*see* OR) functions, which are constructed using metal-oxide semiconductor (MOS) or bipolar circuits. They are used as alternatives to random-access memories (RAM) which use standard logic networks programmed to perform a specific function.

**PLATO** Programmed Logic for Automatic Teaching Operations. A computer language and also a system of computer-aided instruction (CAI).

**PLL** Phase Locked Loop. A phase comparator and a voltage controlled oscillator (VCO) linked so that the phase oscillator frequency can precisely monitor that of a phase-modulated signal or applied frequency. If they differ in phase a control voltage (V) is developed which, when filtered, changes the frequency to reduce error.

**PL/M** Programming Language for Microprocessor. A high-level programming language that permits users to employ a notation with which they are already familiar. It was developed by Integrated Electronics (INTEL), for use on its microcomputer systems, from a dialect of programming language one (PL/1).

**PLM** Pulse Length Modulation. A form of pulse modulation (PM,2) in which the duration of each pulse in a pulse train is varied according to the instantaneous value of the modulating wave. The modulating signal can vary when the leading edge, the trailing edge, or both edges of each pulse occurs.

**PLO** Phase Locked Oscillator. A phase locked loop (PLL) circuit that is used for precise data recovery in floppy disk drive controllers (FDC). It stabilizes the separated data and clock (CLK) binary digits (BITs).

**PLP** Presentation Level Protocol. An agreed viewdata standard between Canada and the USA for the whole of North America.

**PM** (1) Phase Modulation. A method of modifying a sine wave signal to make it carry information, in which the carrier has its phase changed according to the information to be transmitted. In operation the angle relative to an unmodulated carrier angle is varied in accordance with the instantaneous value of the amplitude of the modulating signal. For digital transmission, 2,

4, or 8 different phases can be employed.

(2) Pulse Modulation. A method of transmitting information by modulating a pulsed or intermittent carrier. Pulse width, count, position, phase and/or amplitude may be the varied characteristic.

(3) Post Mortem. (a) An analysis undertaken of an event or operation after the completion of that event or operation. (b) In a computer system, a check routine that provides information concerning the contents of all or a particular section of storage after a problem has ceased to exist. It assists in locating a program error or machine malfunction.

**PMBX**   Private Manual Branch eXchange. A private telephone exchange located on a subscriber's premises, operated manually and providing access to the public telephone network.

**PMD**   Post Mortem Dump. The transfer of the contents of the computer memory instantaneously to a peripheral unit, performed at the end of a machine run. It is used to provide information for locating, diagnosing, and correcting errors.

**PMOS**   P-type (or channel) Metal-Oxide Semiconductor. The earliest of the large-scale integration (LSI) chips (CHIPs). It provides excellent component density, but is slower in performance than N-channel metal-oxide semiconductor (NMOS) devices. The operating speed is of a few megahertz, the device requires two power supplies and is not transistor-transistor logic (TTL) compatible.

**pn** or **p-n**   P-doped N-doped junction. A junction or transition (depletion layer) between a p-doped semiconductor (p-type)

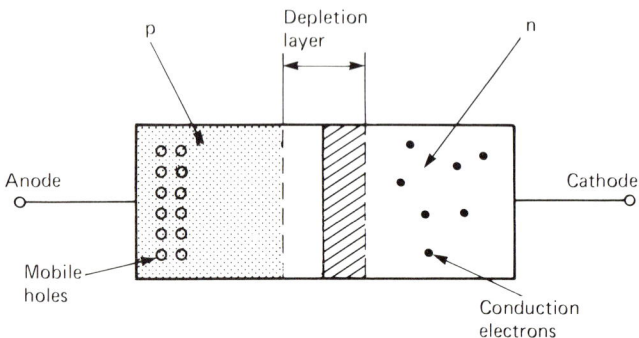

*Figure P.2*   Schematic structure of a p-n semiconductor junction

and an n-doped semiconductor (n-type) having rectifier properties. It allows only a small current to flow due to free electrons and holes being produced near the junction by thermal agitation. *See Figure P.2.*

**PN** Punch oNly. The function of a paper tape punch (PTP,2) which is a device that punches blank paper tape, but is unable to sense or translate the holes.

**PNA** Project Network Analysis. *See* (PNT).

**pnip** P-doped N-doped Intrinsic P-doped transistor. A transistor that is constructed with an intrinsic (undoped) layer of material between the base and collector in order to extend the high-frequency range.

**pnp** or **p-n-p** P-doped N-doped P-doped transistor. Consisting of two p-doped regions (p-type) with an n-doped region (n-type) separating them, this type of transistor is characterized by the flow of 'holes' as majority carriers rather than the more mobile electrons.

**pnpn** or **p-n-p-n** P-doped N-doped P-doped N-doped semiconductor device. Fabricated in silicon, a device that consists of alternating p-doped semiconductor material (p-type) and n-doped semiconductor material (n-type) with at least three p-n junctions (p-n). A basic pnpn structure with no gate electrodes is termed a four-layer diode (D,5) or Shockley diode. It has a bistable current-voltage characteristic and is used for power-switching purposes, as shown in *Figure P.3*.

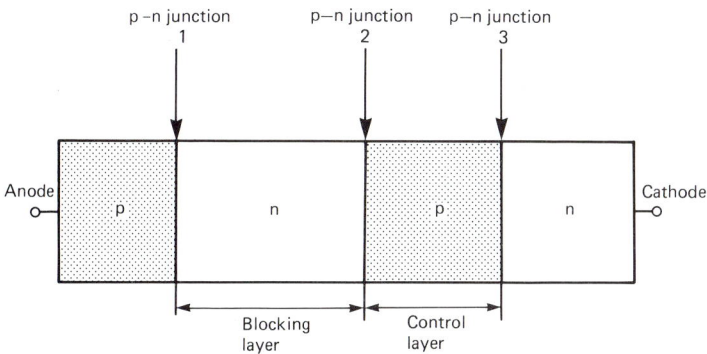

*Figure P.3* A basic pnpn structure with no gate electrodes; known as a four-layer diode, or Shockley diode

**PNT** Project Network Techniques. A group of techniques that considers the logical inter-relationship of all project activities. It is used for the description, analysis, planning and control of projects. The group includes techniques concerned with time, resources, costs and other uncertain factors.

**PO** Post Office. The British Post Office, formerly known as the General Post Office (GPO). The PO has set many standards for the production of electronic devices and equipment.

**POA** Point Of Action. Describes displays (D,6) and readouts sited where processes such as manufacturing actually take place. They are characterized by housings which can stand environmental hazards such as impact and the ingress of dust and water. They have bright wide-angle characters that are often laid out in a single line, so that messages are received by operators without the need to look closely at visual display units (VDUs).

**POKE** A command statement used in Beginner's All-purpose Symbolic Instruction Code (BASIC). This self-modifying feature will cause a computer or microcomputer to change a section of its own program in a manner the programmer commands.

**POL** Problem Orientated Language. A computer language designed for the solutions of a given class of problems. Although elaborate it should be easily understood and usually allows the solution to a problem to be expressed in the same terms as the problem.

**POLISH** POLISH notation. Designed in 1929 by a Polish logician, Jan Lukasiewicz, this is a representative notation in which each operator precedes the operand upon which it operates.

**POLYSILICON** POLYcrystalline SILICON. A polycrystalline form of silicon generally used as the gate electrode in silicon gate metal-oxide semiconductors (MOS), integrated circuits (IC) and charge-coupled devices (CCD). In this particular application the silicon exhibits metallic properties, as it degenerates due to high doping concentrations.

**POP** A colloquial expression for the retrieval of information from a push down stack, in order to place it in a particular memory location or register (as opposed to entering information which is pushed. *See PUSH).*

**POS** Point-Of-Sale. Refers to computer systems that are employed in many parts of retail operations. They can be used for credit authorization, credit verification, stock control or for the transfer of funds.

**POSH** Permuted On Subject Headings. An index that lists all the subject headings of a document's title so that each heading appears, in turn, as the leading word followed by the remaining words.

**POT** POTentiometer. Normally a circular or disk-type of resistor that has two fixed terminals and another third terminal which is connected to a movable contact arm. It is most commonly found as a volume or tone control on audio and television (TV) equipment.

170

**POTS** Plain Old Telephone Service. A term used, mainly in the USA, by the telephone industry for a conventional telephone service.

**PP** Peripheral Processor. Used in association with one or more pieces of computer peripheral equipment (PERIPHERAL), this processor helps the peripheral to function semi-autonomously, not under the control of the central processor unit (CPU).

**PPI** Plan Position Indicator. A type of presentation used in a radar receiver. The signal received from the target is shown as a bright spot on a screen. The distance and bearing of the target are given by the polar co-ordinates of the spot with respect to the centre of the screen.

(2) Programmable Peripheral Interface adaptor. A peripheral interface adaptor that can provide control over the connection of parallel orientated peripherals to various microprocessors (MP).

**PPM** (1) Pulse Position Modulation. A form of pulse modulation (PM) in which the positions, in time, of pulses are varied without their duration being modified.

(2) Peak Programme Meter. An instrument specifically designed to measure the peak values of audio signals within a predetermined period. It is able to respond very quickly to increases in audio signal amplitude, thus enabling accurate measurement of peaks, but takes appreciably longer for indicators to return to zero.

(3) Period Permanent Magnet. A magnet constructed from alternate permanent magnetic rings and soft-iron pole pieces that is able to provide the necessary axial magnetic field required in travelling-wave tubes and other similar devices.

**pps** (1) Pulses Per Second. The number of pulses which occur per second experienced at a particular point in a transmission medium of free space, or at a point in a circuit (CCT), device or system.

(2) Parallel Processing System. A system including a central processor unit (CPU), random-access memories (RAMs), read-only memories (ROMs), and clock (CLK) circuits, with circuits (CCT) that can be used to build equipment requirements into digital (D,4) data processing (DP,1) equipment. Such systems can be found in calculators, general-purpose data processors, process controllers and cash registers.

**PPSU** Programmable Power Supply Unit. A particular type of laboratory power supply (LabPS) unit that has an internal memory which can be programmed with the value of voltages (V,2), each with an associated current-limit and dwell time. A typical unit has an IEEE-488 (*see* IEEE-488) interface and a memory that once loaded can be stepped through, run continuously, or run in a single slot mode. Such features make PPSU devices ideal for use with automatic test equipment (ATE).

**PR** PRefix. A number, syllable or word, placed at the beginning of another number, syllable or word, which together form a compound that has a precise value or meaning.

**PRE** PREfix. *See* PR.

**PREAMP** PREAMPlifier. An amplifier used to amplify received signals before they are fed into the rest of the system. It is used in a sound reproduction system such as a radio receiver, in an antenna system, and in television (TV) cameras.

**PRECIS** (1) PREserved Context Index System. A system of indexing information in which various groups or strings of items are arranged with respect to their linguistic function. This permits a computer to use the information strings so that the user can find a term which is of primary interest.
(2) PRE-Coordinate Index System. A system of indexing documents in which terms are combined as the index is being compiled to provide the entries.

**PRISM** Personal Records Information System Management. A computerized personal information retrieval system (PIRS) used by the UK Civil Service.

**PRF** Pulse Repetition Frequency. The number of pulses measured during a particular time interval, which are independent of that interval. The term is normally used for a regular train of pulses, i.e. the number of pulses per unit time is constant. With reference to computers, the PRF is the number of electric pulses experienced in a particular unit of time, measured at one point within the system.

**PROM** Programmable (or Programmed) Read-Only Memory. Any type of memory device that is not recorded during manufacture, but requires some form of physical operation to be programmed. Special processes can allow some PROMs to be erased and reprogrammed (*see* UV PROM). They are used to form read-mostly memories (RMM) of which the most common type is the floating-gate PROM, which employs the metal-oxide semiconductor (MOS) as its basic memory cell. The cell has two gate electrodes, one above the other, separated by a layer of silicon dioxide. The lower gate is completely surrounded and therefore is floating.

**PRR** Pulse Repetition Rate. The number of pulses that occur in a unit time experienced at a particular point in a transmission medium of free space, or at a point in a circuit (CCT), device or system. The rate is normally expressed in pulses per second (pps).

**PRT** Production Run Tape. A tape containing a verified and checked schedule for a production run, and able to run on a variety of computers.

(2) Program Reference Table. A particular memory-storage area, used to store operands, array references, file references and references to segments of a computer program. It allows programs to be independent of the memory locations actually occupied by parts of the program and data.

**ps** PicoSecond. One millionth of a second (s), that is $0.000\,000\,000\,001$ or $10^{-12}$ second.

**PSD** Post Sending Delay. A time interval that occurs between the finish of dialling on a telephone and the reception of a return signal, which could be a ringing tone or a busy tone (BT,2).

**PSE** Packet-Switching Exchange. A group of network nodes (intersections of a number of networks), at one location. The splitting of information into packets for transmission is often done at the PSE. *See* PSS,1, packet-switching system.

**PSK** Phase-Shift Keying. A process whereby the phase of a modulated carrier-wave is changed between a specified set of predetermined values.

**PSN** Public Switched Network. A switching system made available by a common carrier for circuit switching many subscribers. An example in the UK is British Telecom (BT,1).

**PSS** (1) Packet-Switching System. A data transmission system that provides for the transfer of discrete groups or packets of information along a channel (CHNL(a)), which is occupied only for the duration of the transmission of the packet. The packet itself contains address information and the channel is available for the transfer of other packets, providing communication through the network to other users without the need for a two-way transmission path.

(2) Packet-Switched Service. The packet-switched network commissioned in 1980, and run by the Post Office (PO) in the UK.

(3) Packet-Switch Stream. *See* PSS,1, packet-switching system.

**PST** Pair-Select Ternary. A pseudo-ternary code where pairs of binary digits (BITs) are coded together, making a third value, in such a way that the resultant signal has no long strings of zeros. A ternary code has three possible values, for example $+1$ volt (V), O and $-1$ V.

**PSTN** Public Switched Telephone Network. *See* PSN.

**PSU** (1) Power Supply Unit. All electronic devices and circuits require some form of external power source. The voltages required range from 1.5 volts (V) to 25 kV or higher, and the currents can range from a few microamps (mA) to several tens of amps (A). PSUs can be produced to cover all these requirements. Most electronic circuits run off the 240 volts (V), 50 hertz (Hz) supply via a PSU, the simplest form of which consists of a transformer that produces alternating current (AC) at a lower voltage (V), which is in turn rectified by a rectifier circuit to give off direct current (DC). The high ripple content of the rectifier's output (O,2) is removed by a smoothing circuit to provide a stable supply. Also some form of stabilization is provided as the DC output has to remain substantially constant free from the effects of changes in load current, mains input and temperature. A simplified schematic diagram is shown in *Figure P.4.*

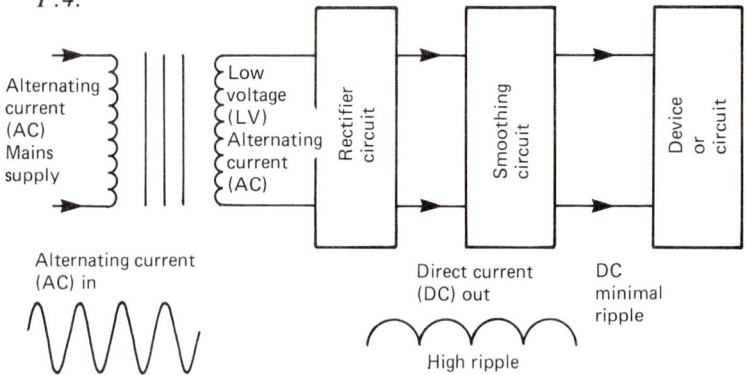

*Figure P.4* A simplified schematic diagram of a power-supply unit (PSU)

(2) Program Storage Unit. A chip (CHIP) that principally provides for the storage of programmed instructions and various nonvolatile data constants of a program. It is able to be interfaced directly with a central processor unit (CPU) without the need of buffer circuits.

**PSW** (1) Processor Status Word. A register held in a central processor unit (CPU) containing the status of the most recent instruction execution result. It can record error and exception conditions that are able to be dealt with by the operating system (OS).

(2) Program Status Word. A sector of main memory storage that is used as a computer system status indicator in relationship to a specific program.

**Pt** PlaTinum. An extremely stable noncorroding metal with the atomic number 78. It is used as a conductor or contact at high temperatures or where chemical attack is expected.

**PT** Paper Tape. A strip of paper normally 25 mm wide and up to 300 m in length on which data is recorded in the form of holes arranged in groups of binary format along the tape's length. The tape is divided into columns which run the length of the tape and channels which extend across the tape width, the number of which depends on the system used.

**PTC** Positive Temperature Coefficient. The attribute of a semiconductor material to have a resistance that increases with an increase in temperature. PTC semiconductor materials are normally employed in thermally sensitive resistors (THERMISTOR).

**PTFE** PolyTetraFluoroEthylene. A thermoplastic that can be used over a wide range of temperature and is very resistant to moisture. It has excellent electrical insulating properties and can be moulded or shaped into many forms that are used in the electrical industry.

**PTM** Pulse Time Modulation. A term, used generally, to describe various methods of modulation in which a pulse carrier that is time-related varies according to the instantaneous value of a modulated signal.

**PTP** (1) Point-To-Point. The transmission of data between two points. It can also relate to the single physical line, switched or non-switched, that connects a remote terminal to a computer. (2) Paper-Tape Punch. An output (O,2) device that punches or encodes machine-readable output data in the form of holes into paper tape (PT). It has been a very popular device for producing physical data storage. (*See* PTR).

**PTR** Paper-Tape Reader. A popular low cost input device that senses and translates the holes or information provided in a paper tape (PT) into machine code. The use of a paper-tape punch (PTP) and a PTR combination has now been generally superseded by magnetic tape (MAGTAPE) cassette units.

**PTS** Program Test System. Capable of automatically verifying and checking programs, this specific system is able to assist in the organization of a production run by providing diagnostic information.

**PTT** (1) Postal, Telegraph and Telephone authority. The particular national authority that is in overall command of posts and telecommunications within a particular country. In the UK the General Post Office used to be the PTT, until it was replaced by two authorities, the Post Office (PO) and British Telecom (BT,1). In the USA there is no corresponding body, although the Federal Communications Commission (FCC) regulates these matters.

175

(2) Push-To-Talk. A switch that is depressed during transmission and released during reception, normally on the handsets of two way radio communication devices such as citizen band radios (CB), ship-to-shore radios and police communicators. Also known as 'pressed switch'.

**p-type**   An extrinsic semiconductor containing a higher density of mobile holes (Valence band electrons) than of conduction electrons, which means the holes are the majority carriers, *see Figure P.5.* Compare n-type semiconductor.

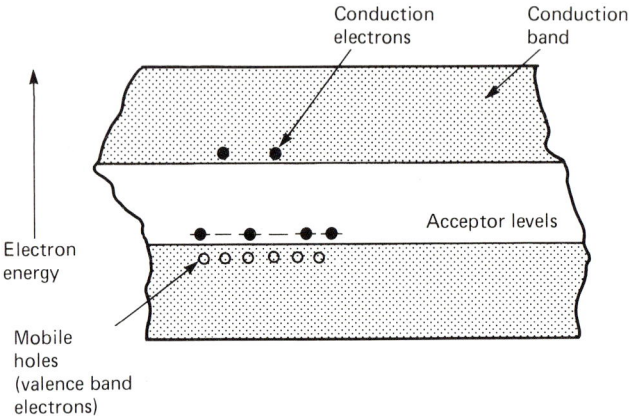

*Figure P.5*   Energy bands of a p-type semiconductor

**PU**   Per-Unit. In the per-unit system quantities are expressed in decimal fractions of unity (one) rather than as parts per 100 (per cent). Advantage is gained in that quantities expressed in per-unit values can be more easily multiplied and divided, making calculation simpler.

**PUFF**   A colloquial expression used for picofarad, it is derived from the markings of 'pF' on capacitors representing picofarads.

**PUFFT**   Purdue University Fast ForTran. A particular version of the formula translator (FORTRAN) computer language.

**PUSH**   An instruction that is used to deposit a word on the top of a stack or push-down list (as opposed to POP).

**P-V**   Peak-to-Valley value. The comparison or difference between the greatest instantaneous value (peak value) and the least instantaneous value (valley value) of a time-dependent quantity.

**PVC**   PolyVinylChloride. A thermoplastic that is available in a wide range of bright colours and can be transparent or opaque. Being resistant to water, corrosion and mechanical abrasion, it

176

is used widely in the electrical industry for sheating and insulation of extruded cable.

**PVDF** PolyVinylDene Fluoride. A semicrystalline, high molecular weight polymer of the repeat unit, $CM_2$-$CF_2$ that is used as transducer film material. It is approximately 50% crystalline and 50% amorphous and has a high piezo-electric response.

**PVT** Page View Terminal. A visual display unit (VDU) that allows the user to display (D,6) pages of text, one page at a time.

**PW** Private Wire. A specific channel (CHNL(d)) along with the necessary equipment, that is intended for the exclusive use of a particular subscriber. Application of such a line can include a telephone link, a facsimile (FAX) or telecommunications link.

**PWB** Printed Wire Board. A conductive pattern that has been formed, by etching or plating, on the surface of an insulating baseboard. *See* PCB, printed circuit board.

**P-wire** Private WIRE. A wire over which automatic switching functions of guarding, holding and releasing are controlled and also in some systems over which metering may be achieved.

**PWM** Pulse Width Modulation. Pulse modulation (PM,2) in which the duration of each pulse in a pulse train is varied according to the instantaneous value of the modulating wave. The modulating signal can vary when the leading edge, the trailing edge or both edges of each pulse occurs.

**PX** Private eXchange. A telephone exchange that serves a particular organization and has no means of connection with the public telephone network.

**PZT** PieZoelectric Transducer. A crystal that produces a voltage (V) when stressed by compression, expansion or torsion. The voltage is only produced when the applied stress is changing.

# Q

**Q** (1) Queue. Arranged in sequence, a collection of items that has two ends which are the head and the tail. New items are added to the tail. Items can be removed from either the head or the tail.

(2) A register used as an accumulator extension. This is necessary for efficient multiple-divide computer programming.

(3) An address, or source location in internal storage in some forms of computer equipment from which data can be transferred.

(4) A comparison test made between two or more units of quantitative data to discover if they are equal or unequal.

*See also* Q-axis, Q band, Q-component, Q-factor, Q-meter, Q-signal, Q-switch.

**QA** Quality Assurance. Methods employed in a systematic manner to provide the necessary confidence that an item itself has been tested or is constructed of components which have been tested and will perform satisfactorily.

**QAGC** Quite Automatic Gain Control. A delayed automatic gain control (AGC) combined with a signal suppressor in such a way that the output (O,2) of a receiver is suppressed for inputs too weak to operate the AGC.

**QAM** (1) Quadrature Amplitude Modulation. A high-speed modulator/demodulator (MODEM) modulation technique used in digital (D,4) transmission where two sinusoidal carrier waves having a phase difference of 90° between them are amplitude modulated (AM) by independent signals and the resulting components are summed to produce the output (O,2) signals. This system is used in colour television (TV) transmission.

(2) Queued Access Method. A term applied to any data access method that is able to synchronize the data transfer between the user program and input/output (I/O) devices and reduce input/output delays.

**Q-axis** The chrominance axis that corresponds to the coarse chrominance primary in the National Television System Committee (NTSC) colour television transmission system. Also known as narrow-band axis.

**Q band** Microwave frequencies. Q-band microwave ($\mu$W) frequencies occur between 36 and 46 gigahertz (GHz) and form part of a larger continuous range of frequencies covering 0.225 to 100 gigahertz which are usually subdivided into bands designated by letters. Q-band frequencies have wavelengths of between 0.834 and 0.652 cm.

**QC** Quality Control. Workmanship, processes and materials that have been monitored by an organized and formal process to produce consistent and uniform products. The process can involve the keeping of evaluation records, batch logs, and error summary reports.

**QCB** Queue Control Block. A memory storage location that holds information in a condensed form, which is used to regulate the sequence in which entries in a list waiting for service are processed.

**Q-component** Resulting from the modulation of a sub-carrier by the Q-signal, this is the component of the chrominance signal in the National Television System Committee (NTSC) colour television broadcasting system.

178

**QED** Quick text EDitor. A high-speed system that provides the user with a flexible source text generation system. Source statements are entered which can be output (O,2), statements added, or groups of statements altered.

**Q-factor** Quality-FACTOR. A measure of the ratio of energy stored in a system, circuit, component or device in comparison to the energy dissipated. The energy dissipation could be due to resistance, electromagnetic radiation or other factors.

**QIL** Quad-In-Line. A dual-in-line package (DIP) with four similar functional circuits or units (two rows of pins on each side). *See* QUIP.

**Q-meter** An instrument to measure the Q-factor (*see* Q-factor) of a tuned circuit or of circuit components.

**Q-point** Quiescent POINT. The area on the characteristic curve of an active device, such as a transistor, when the device has no input signal.

**QPL** Qualified Products List. A list used in the USA containing products that are considered to be of high reliability and are qualified for military (MIL,2) use.

**QPP** Quiescent Push-Pull. A circuit that operates in a push-pull mode using two pentodes biased near anode-current cut off. The circuit was once often used in battery-operated equipment.

**QSAM** Queued Sequential Access Method. A type of the basic sequential access method (BSAM), which uses blocks formed from queues (Q,1) of input data which are waiting for processing, or output (O,2) data blocks which have been processed and are waiting for transfer to output devices or other auxiliary storage.

**Q-signal** The signal that represents the chrominance information along the green-magenta axis in the National Television System Committee (NTSC) colour television transmission system.

**Q-switch** A device that prevents pulsed laser emission being achieved until a certain required energy level is reached in the lasing material. It provides shorter and more intense pulses at a higher repetition rate than could be reached by pulsing the lasing material.

**QTAM** Queued Telecommunications Access Method. A method of transferring data between main computer storage and remote terminals. It uses an application program with specific macro instruction that requests the transfer of data, which is performed by a message control program that also synchronizes the transfer, which in turn eliminates delays for input/output (I/O) operations. The technique is employed for data collection, message switching and other teleprocessing uses. It provides the capabilities of basic telecommunications access

*Figure Q.1* A QWERTY keyboard (*courtesy* Alphameric Keyboards Ltd)

180

method (BTAM) as well as the ability of queued messages on direct-access storage devices.

**QUICKTRAN** QUICK forTRAN. A development of the formula translation (FORTRAN) programming language designed primarily for running from a remote terminal.

**QUIP** QUad-In-line-Package. Having two rows of pins along each longitudinal edge of its package, rather than one row as in a dual-line package (DIP), this 64-lead integrated circuit (IC) package was developed jointly by Integrated Electronics (INTEL) and 3M.

**QWERTY** A standard (English-language) computer terminal keyboard (KB) or typewriter keyboard on which the keys are arranged such that the top row below the numerals begins with the letters Q, W, E, R, T and Y, reading left to right, as shown in *Figure Q.1*. Mainly used in the UK and USA as well as other English-speaking countries. (*See* AZERTY and QWERTZ.)

**QWERTZ** A German-language computer or typewriter keyboard with the keys arranged so that the top row below the numerals begins with the letters Q, W, E, R, T and Z, reading left to right, as shown in *Figure Q.2*. It normally contains some additional letters peculiar to the German language, as well as instructional keys notated in German. It is used mainly in German-speaking countries. (*See* AZERTY and QWERTY.)

# R

**R** (1) Resistance. Having the unit ohm, this is a measure of the tendency of a particular material to resist the passage of an electric current (I) and to dissipate electrical energy into heat energy. It is the ratio of applied potential difference (PD) across a conductor to the current flowing through it.

(2) Resistor. The symbol R is used to represent resistor on parts lists and to annotate schematic diagrams.

(3) Reluctance. Magnetic resistance, the ratio of magnetomotive force to the total magnetic flux having the unit henry $^{-1}$.

(4) Rontgen. A measurement of the effect of ionization. It can be defined as the quantity of gamma- or X-rays such that associated emissions every 1.293 mg of air produce ions in the air that carry 1 electrostatic unit of charge.

**RACE** Random-Access Computer Equipment. Computer equipment in which information can be introduced and stored, where any location can be read from, or written into, in a random access method. It is used where the information to be obtained is unrelated to the location of previous information.

**RAD** Rapid- (or Random-) Access Disk. A memory storage

*Figure Q.2* A QWERTZ keyboard (*courtesy* Alphameric Keyboards Ltd)

182

device often a magnetic disk that provides rapid access to a memory location, and retrieval of data, a procedure that is often synonymous with random access or direct access. It is contrasted with sequential access in that it does not depend upon the access of preceding data.

(2) Radiation Absorbed Dose. The base unit of the amount of radiation absorbed which results in the ionization of the material onto which it falls or strikes the surface of. A dose of 1 RAD equals the absorption of 100 ergs of radiation energy per gram of absorbed material.

**R&D** Research AND Development. The application of scientific and technical activities directed towards the development of new products systems or services.

**RADA** Random-Access Discrete Address. A particular location in a random access memory (RAM).

**RADAR** RAdio Direction And Ranging. Using reflected radio waves of microwave ($\mu$W) frequency, a RADAR system is able to locate distant objects. Current sophisticated radar systems can provide information concerning both stationary and moving objects and can be used for navigation and guidance of ships, aircraft and other vehicles and systems.

**RADIR** Random-Access Document Indexing and Retrieval. A system for creating an index of documents, in which access to data to be obtained is unrelated to the location of previously accessed information.

**RALU** Register and Arithmetic Logic Unit (or Register-equipped Arithmetic Logic Unit). An assembly of logic elements including accumulators, stack and arithmetic logic units (ALU) which together provide both storage and processing functions. All the elements of the unit are married by a set of buses.

**RAM** Random-Access Memory. A direct-access memory device that is used where it is required to change the binary digit (BIT) location during the routine operation of the memory. It is a mass storage device designed so that the location of stored data is independent of content and any location may be directly accessed without having to work through from the beginning. The fast access to any storage location is provided by means of vertical and horizontal co-ordinates and information is written in or out using the same procedure. Sometimes also known as a 'random-access/read-write memory'.

**RAMIS** Random-Access Management Information System. A system providing information in response to users' requests in which the computer files store management data in a random-access memory (RAM).

**RAS**   Row Address Strobe. Used in dynamic random-access memories (DRAM) this signal is employed to reduce the number of pins necessary, by multiplexing the address.

**RATFOR**   RATional FORtran. A structured development of formula translator (FORTRAN) which is translated or compiled into standard FORTRAN by a processor.

**RAYNET**   Radio Amateurs emergencY NETwork. An organization set up to provide assistance to the authorities during disaster relief operations, by providing a means of communication.

**RAX**   (1) Remote ACCess. The use of a terminal that is capable of providing access to a computer from a point that is physically distant to the computer.
(2) Rural Automatic eXchange. An automatic telephone exchange located in the countryside. Due to its position it deals with a lower average number of telephone calls than an automatic exchange located in a town or city.

**RB**   Return to Bias. A particular mode of recording where the state of the medium used for recording alters from a bias state to another state, and then returns to record a binary 1.

**RBE**   Remote Batch Entry. The transfer of a group of records to be processed as a single item, over a distance by means of a communications link.

**R-C**   Resistance-Capacitance. A prefix used to describe devices or circuits where the use of resistance (R,1) and capacitance (C,5) is necessary for their operation, or that employ resistance-capacitance coupling.

**RC**   Regional Centre. Class 1 of five class numbers assigned to offices according to their function in the American direct distance dialling (DDD) telephone network. Any one centre can handle telephone call traffic from one, two or more centres lower in the hierarchy, Class 1 being the top classification.

**RCCB**   Residual Current Circuit-Breaker. A term used in switching terminology that is superseding earth leakage contact-breaker (ELCB) and conforms to current regulations. RCCBs can provide user's with protection against serious electrocution either from appliance or cable by tripping out well before a serious shock can be received, typically in 30 milliseconds and at 30 milliamps.

**RCTL**   Resistor-Capacitor-Transistor Logic. The logic elements of a circuit or system which are similar to those of a resistor-transistor logic (RTL) circuit, except that capacitors (C,6) are provided to enhance switching speeds.

**RCV**   ReCeiVe. The process of obtaining a transmitted signal, as contrasted with sending or transmission.

**RD** ReaDer only. The function of a paper-tape reader (PTR), which is a device that senses and translates the holes or information on a paper tape (PT) into computer machine code, but is unable to punch holes or put information on blank tape.

**RDF** Repeater Distribution Frame. Housed in a telecommunication repeater station, this frame provides interconnection for various items of apparatus or equipment which include signalling units, amplifiers (AMP) and transformers.

**RDOS** Real-time Disk Operating System. A disk operating system (DOS) that operates in real time (RT,1).

**RDY** ReaDy. The control signal, used in some systems, for slow memory to indicate if data is valid.

**READ** (1) Real-time Electronic Access and Display. The acquisition of information from a storage device or a peripheral unit, which is produced on a visual display unit (VDU). The whole operation taking place in real time (RT,1).

(2) Remote Electrical Alphanumeric Display. A visual display, normally on a visual display unit (VDU), of numerical and alphabetical details, that is provided at a location physically distant from the computer.

**REALCOM** REAL-time COMmunications. A communication system in which response occurs at the same rate as that of the data input.

**RECOL** REtrieval COmmand Language. A computer language employed to interrogate the store of data held on the computer files within an information retrieval system.

**REED** Restricted Edge Emitting Diode. A light-emitting diode (LED) that emits light over only a small area of one edge. This region of restricted light emission, when used with integrated optical circuits (IOC,2) and optical fibres (OF), improves coupling efficiency.

**REGEN** REGENerative repeater. A device used to amplify, retime and retransmit received signal impulses at their original strength or level.

**REM** REcognition Memory. A simplified associative memory designed to interface simply with ordinary computers.

**REMSTAR** Remote Electronic Microfilm in Storage Transmission And Retrieval. A computer output microfilm (COM) station that is physically distant from the main system. It is able to provide necessary distribution and storage functions.

**RES** REStore. (a) To return a device or circuit to its initial state. (b) To return a counter or timer to zero. (c) To return a computer word to its initial or preselected value. (d) To return a register, trigger or error correction signal to zero or to a specific initial condition. (e) To place a zero in a binary cell.

**RETD** Red Especial de Transmission de Datos. The Spanish postal telegraph and telephone authority (PTT) packet-switching network.

**REW** REWind. To return a magnetic tape (MAGTAPE), film or paper tape (PT) to its beginning, or some other initial data point.

**REX** Real-time EXecutive. A sequence of operations that a computer is to perform, in which the computer responds at the same rate as that of the data input.

**RF** Radio Frequency. A frequency of electromagnetic radiation or alternating current (AC) within the range of 3 kilohertz (kHz) to 300 gigahertz (GHz). By international agreement radio frequencies occupy the bands numbered between 5 and 11 (inclusive) and have wave lengths 100 km and 1 mm. Each band extends from $0.3 \times 10^N$ hertz (Hz) to $3 \times 10^N$ Hz where $N$ is the specific band number.

**RFC** Request For Connection. Required to achieve a connection, this command is used in the formalized method by which two parties communicate in an advanced research projects agency network (ARPANET) packet-switched computer network.

**RFI** Radio Frequency Interference. The distortion or impairment of a desired transmission or signal brought about by radio frequency (RF) disturbances such as atmospheric noise.

**RG** Release Guard. The protection of a piece of telephone equipment from seizure (i.e. calls to be established) or attempted use whilst it is returning a line from a busy state to a free state.

**RGB** Red, Green, Blue signals employed in colour television (TV) transmission, these three different signals each represent one of the three primary colours. Also known as 'tristimulus' signals.

**RGS** Release Guard Signal. This signal protects a telephone circuit or channel from seizure (i.e. calls being established) at its outgoing end whilst equipment at its incoming end is returning from a busy to a free state.

**RI** Radio Interference. A phenomena experienced when an unwanted radio signal or radio frequency (RF) disturbance such as 'noise' or 'atmospherics' interferes with a wanted or required radio signal.

**RICASIP** Research Information Centre and Advisory Service on Information Processing. An organization in the USA sponsored by the National Science Foundation and the National Bureau of Standards (NBS). Its aims are to provide aid and assistance concerning information processing.

**RIG** A slang expression used to describe transmitter or transmitter receiver (TRANSCEIVER) equipment.

**RIT** Receiver Incremental Tuning. A control that permits the frequency received in a transmitter/receiver (TRANSCEIVER) to be carried (typically ±5 kilohertz (kHz)) on one side or the other of the transmitted frequency.

**RITA** Rand Intelligent Terminal Agent. A RITA system permits a terminal simple access to a computer network.

**RJE** Remote Job Entry. The action of a remote device providing input of job information to a main computer system.

**RLD** ReLocation Dictionary. A section of a program that contains the information necessary to alter the label identifying the location of where it is stored, after it has been placed in a new location, i.e. relocated.

**RMM** Read-Mostly Memory. A memory device used where long-term storage of information is required. Thus, stored information is altered very infrequently, so allowing for a device that has the write time much greater than the access time. (*See* ROM).

**RMS** Recovery Management Support. When operations fail due to some hardware error, this facility gathers information concerning hardware reliability and allows the operations to be attempted again.

(2) Root-Mean-Square. The square root of the average of the square of a variable. It is usually employed to express the effective value of an alternating current (AC). The RMS value of an alternating current is the square root of the average of the square of the current taken throughout one period. This is comparable to a direct current (DC) in terms of heating effect, etc.

**RMW** Read-Modify-Write. A refresh cycle used in the operation of a random-access memory (RAM).

**R/NAV** Radio NAVigation system. A computerized avionic system for small military and commercial aircraft, designed to work with conventional cockpit instruments. It enables a pilot to fly along any course without manual calculations or reference to maps and is able to provide dead reckoning navigation when temporarily outside the range of a radio beacon.

**RO** (1) Read Only. Data access that allows the data to be examined or read, but does not permit any modifications to be made.

(2) Receive Only. A printer unit in data processing (DP) or a radio receiver in telecommunications (TELECOMMS) that can only receive messages and is unable to generate them.

**ROM** Read-Only Memory. A computer memory with a permanent set of information written into it, whose contents under normal conditions are not intended to be alterable by instructions. There are three groups of ROMs: masked programmed, fusible link (FL) and alterable. The mask programmed ROM is programmed by a mask pattern, as a part of its final manufacturing process. The fusible link is permanently programmed by electrical pulse or mechanical means. The alterable device undergoes changes which are induced electrically.

**ROOST**   Rapid Optical Ocean Surveillance Testbed. A research and development (R&D) programme to refine optical signal processing techniques for use in ocean surveillance, sponsored by the US Navy and the US Defense Advanced Research Projects Agency (DARPA). It is a hardware-orientated engineering programme to test the concept that successful integration of advanced optical, digital and display technology can increase the speed with which acoustic data can be obtained in ocean environments.

**ROPP**   Receive-Only Page Printer. Used at communications stations that are unable to generate messages, this type of teleprinter unit has a printer only.

**RP**   Reader Punch. The function of a paper tape (PT) reader punch. It can punch holes in blank paper tapes and is also capable of sensing and translating holes on a tape into computer machine code.

**RPG**   Report Program Generator. (a) A processing program that provides a convenient method for producing a wide variety of reports. The generator can vary from a listing of a magnetic tape (MAGTAPE) reel to a very formal, calculated and edited table of data provided from a number of inputs. (b) A computer programming language that helps the user by providing facilities for assimilating data, producing printed reports and creating, updating and maintaining files.

**RPM**   Revolutions Per Minute. The total number of, or fraction of, complete turns or cycles made by a rotating device in a minute.

**RPN**   Reverse Polish Notation. A form of logic that allows the user to enter every problem from left to right as it is written or notated. The RPN calculator has no *equals* key, as it uses an *enter* key during operation. For example, to calculate 3 multiplied by 7 the RPN calculator functions thus:

| 3 | ENTER | × | 7 | ENTER | 21 |
|---|---|---|---|---|---|
| PRESS | PRESS | PRESS | PRESS | PRESS | DISPLAY |

on pressing 'enter' the second time the calculation takes place and the answer is displayed.

**RPS**  Revolutions Per Second. The total number of, or fraction of, complete turns or cycles made by a rotating device in a second.

**RRRV**  Rate of Rise of Restriking Voltage. The speed at which the voltage (V) increases across the contacts of an interrupting device. Across an arc initially a very small current flows due to residual ionization. If the RRRV is sufficiently fast the residual ionization increases, and the current is re-established. On the other hand, if the RRRV is too slow, the ionization decays and the current cannot re-establish itself.

**RS**  Record Separator. A particular control character that is used to indicate logical divisions between separate areas of records.

**RS-232C**  A popular and widely used standard for the interconnection of computer terminals, cathode-ray tubes (CRTs), displays (D,6), modulator/demodulators (MODEMs), teletypewriters (TTYs), input/output (I/O) devices, peripheral devices, data terminal equipment (DTE) and data communication equipment (DCE). RS-232C is actually an electrical standard and the suffix C denotes the latest revision.

**RSGB**  Radio Society of Great Britain. The UK national society for radio amateurs which endeavours to promote greater awareness, understanding and use of amateur radio.

**RT**  (1) Real Time. A computer system response which occurs at the same rate as that of the data input. Remote controls or automatic guidance systems by necessity operate in real time.

(2) Radio Telephone. A combined and complete radio receiver and transmitter (TRANSCEIVER) required at one station for two-way communication.

(3) Remote Terminal. A terminal that is physically distant from the process or computer system it is accessing. The link can be made via a cable or a broadcast transmission.

(4) Reperforator/Transmitter. A teletypewriter (TTY) that consists of a reperforator and a tape transmitter which are independent of one another. It performs a relaying function and it can transform the speed of incoming information to a different outgoing speed, and is also used for temporary queueing (Q,1).

**RTBM**  Real-Time Bit Mapping. A process employed to aid production. It can be used to locate defects in the masks used for the diffusion, oxidization and metallization steps in chip (CHIP) manufacture, and can also be used to overcome process problems.

**RTC**  Real-Time Clock. A device that produces readable digits

(DGTs) or periodic signals that allow a computer to recognize elapsed time between events, and also indicates the performance of time initiated or time related processing.

**RTE**   Real-Time Executive. A system of computer software that provides an overall guidance for scheduling, handling and program capabilities.

**RTI**   Radiation Transfer Index. An index given as the 1/10th root of the overall efficiency of a radiation transmission system.

**RTL**   Resistor-Transistor Logic. The logic elements of a circuit or system which comprise resistors and transistors, as shown in *Figure R.1*. There is a family of RTL integrated logic circuits in which the input is through a resistor into the base of an inverting transistor. The basic Not OR (NOR) gate is one.

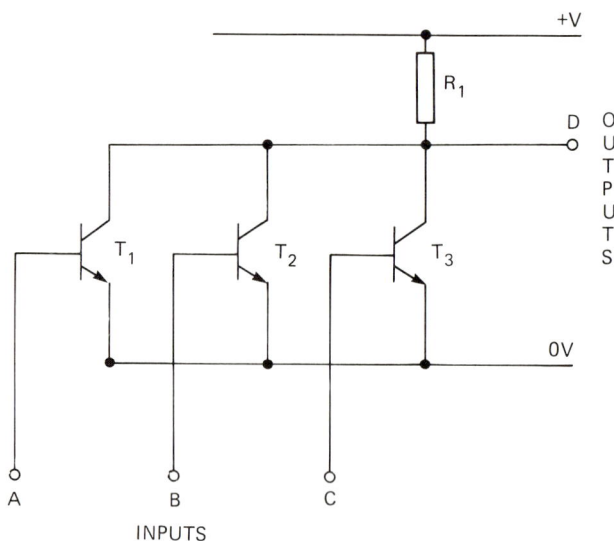

*Figure R.1*   A resistor-transistor logic (RTL) gate

RTLs were the first integrated logic family to be developed but are now little used.

**RTM**   Register Transfer Module. A functional register unit designed for the construction of register logic systems.

**RTOS**   Real-Time Operating System. Software that is required to manage the hardware resources of a computer system, which is capable of real time (RT) task management. The management can include scheduling, real-time event counters and event scheduling.

**RTT**   Regie des Telegraphes et des Telephones. The postal, telegraph and telephone (PTT) authority of Belgium.

190

**RTTY**  Radio TeleTYpewriter. A teletypewriter (TTY) that operates via radio circuits employing frequency-shift keying (FSK) and also audio frequency-shift keying (AFSK).

**RTU**  Remote Terminal Unit. A computer terminal that is physically distant from the main system that it is accessing. The link can be made via a cable or a broadcast transmission.

**RTV**  Room-Temperature Vulcanizing elastomer. A covering, normally made from a silicone rubber compound, that is applied over an entire chip (CHIP) to protect it from its surroundings.

**RUBOUT**  A colloquial expression for 'delete character' (DEL,2).

**R/W**  Read/Write. The function of the small electromagnetic heads which read, record (write) or erase polarized areas on a magnetic surface, usually tape.

**RWD**  ReWinD. A computer program term that is intended as an instruction to rewind a reel of tape.

**R-wire**  Ring WIRE. A wire in a telephone exchange circuit that is connected to the ring-shaped contact of a telephone jack plug or with the corresponding contact of a switchboard jack.

**RWM**  Read/Write Memory. Where the application for a small computer system is fixed, it would have a read-only memory (ROM) program memory. Where it is necessary to change the system application by an operator a RWM is employed. The RWM has provided advancement in small computer system complexity for hardware and programs.

**RX**  Receiver. A unit or circuit in a communications system that converts incoming electrical signals into audible signals. Normally on communications equipment a green panel indicator shows that an RX circuit is in operation.

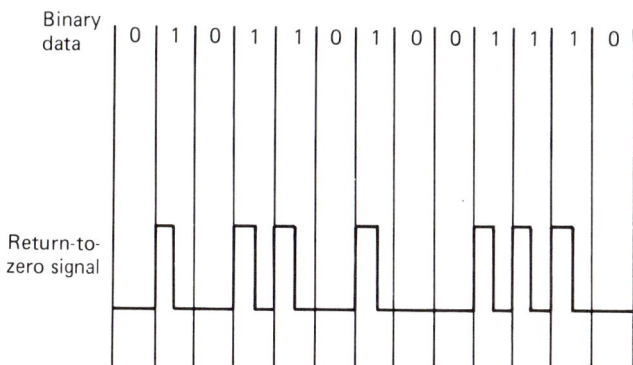

*Figure R.2*  A return-to-zero (RZ) signal

**RZ** Return-to-Zero. A signal presented in a digital (D,4) form having pulses which are not full symbol length, i.e. they are shorter than the spacings between the symbols, as shown in *Figure R.2*. Intervals occur between successive binary 1 symbols, i.e. the signal returns-to-zero.

## S

**s** Second. The SI unit of time, it is defined as the duration of 9 192 631 770 periods of the radiation of the ceasium – 133 atom. Formerly it was defined as 1/86 400th of the average solar day.

**S** Siemens. The SI unit of electrical conductance, susceptance and admittance. A substance possesses a conductance of one siemens if its electrical resistance (R,1) is one ohm.

**SALINET** SAtellite Library Information NETwork. A satellite communications system used to provide library services to the remote areas of Canada.

**SAM** (1) Subsequent Address Message. A message used in common-channel signalling (CCS) following an initial address. It contains one or more additional digits (DGTs) of address information along with or instead of an end-of-pulsing signal.
(2) Serial-Access Memory. A memory-storage device in which the information is entered in sequence. This results in sequential processing of the data.

**SAMICS** Solar Array Manufacturing Industry Costing Standards. A technique that employs a computer program, providing a common ground for the comparison of the economic aspects of several alternative approaches to the production of silicon cells. It can provide some of the best documented and validated costings of the process sequences involved in fabricating solar modules.

**SAT** SATurated mode. The characteristics of operation of a bipolar or field-effect transistor (FET) beyond its pinch-off voltage, i.e. where the pinch-off voltage is less than or equal to the drain voltage. The drain current is independent of the drain voltage in the region.

**SAW** Surface Acoustic Wave. Waves generated on the surface of piezo-electric solids along which they also propagate. They have a velocity that is independent of frequency, but comparable to the velocity of sound through the material. SAW devices are used in bandpass filters.

**S band** Microwave ($\mu$W) frequencies that occur between 1.55 and 5.20 gigahertz (GHz) and form part of a larger continuous range of frequencies covering 0.225 to 100 gigahertz (GHz) which are usually subdivided into bands designated by letters.

192

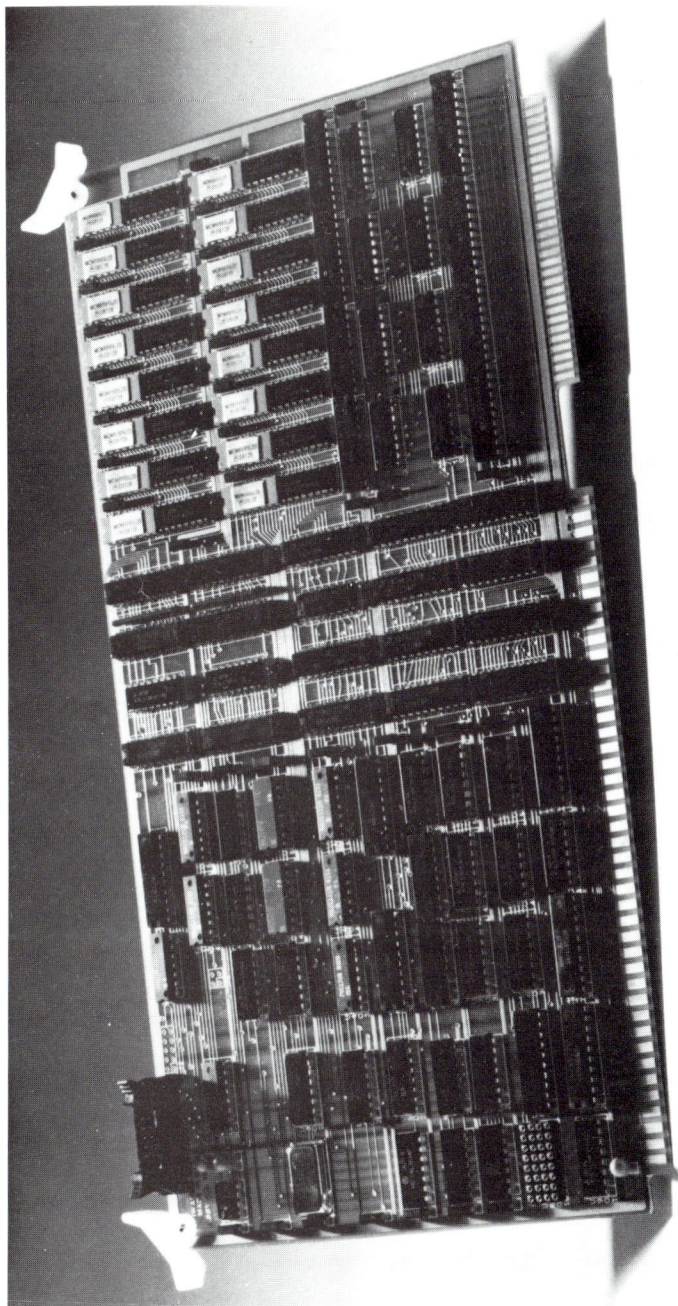

*Figure S.1* A 68000 based multibus compatible single-board computer (SBC) (*courtesy* Data Translation Ltd)

S-band microwave frequencies have wavelengths of between 19.3 and 5.77 cm.

**SBC**  Single-Board Computer. A complete computer that is produced on a single printed circuit board (PCB), *see Figure S.1*, including a central processor unit (CPU), random-access memory (RAM) and all necessary interfaces. It is widely used by original equipment manufacturers (OEM) where an element of intelligent control is required to be built into equipment.

**SBS**  Satellite Business Systems. An organization established to offer satellite communication channels (CHNL(a)) to provide the USA with an electronic mail service.

**SC**  Sectional Centre. Class 2 of five class numbers assigned to offices according to their function in the American direct distance dialling (DDD) telephone network. Any one centre can handle telephone call traffic from one, two or more centres lower in the hierarchy, Class 1 being the top classification.

**SCA**  Steel-Cored Aluminium. Consisting of a core of galvanized steel strands encircled by layers of aluminium, this conductive material combines the strength of steel with the conductive ability of aluminium.

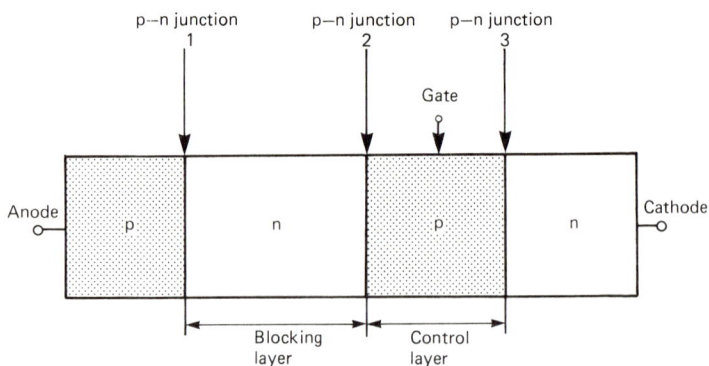

*Figure S.2*  A basic silicon-controlled rectifier (SCR)

**SCARAB**  Submersible Craft Assisting Recovery And Burial. An unmanned submersible craft developed by an international consortium to assist in the recovery, repair and replacement of damaged submarine cables buried in the seabed.

**SCOPE**  OscilloSCOPE. An instrument that is able to provide visible images of single or multiple rapidly-varying electrical quantities. Images can be displayed with respect to time, or with respect to other electrical quantities. The most common type is the cathode-ray oscilloscope (CRO).

**SCR** (1) Silicon-Controlled Rectifier. A pnpn (pnpn) device constructed of a chip (CHIP) of four layers of semiconductor material that form three p-n junctions (p-n). In operation the forward anode-cathode current (I) is controlled by a signal being applied to a third gate electrode. A basic SCR is shown in *Figure S.2. See* CSR.

(2) Sequence-Control Register. A specific storage location or register that holds the location of a specified item in a memory, containing the next instruction to be executed.

(3) Short-Circuit Ratio. The ratio of the field excitation for rated voltage (V) on open circuit, against the field excitation for rated armature current (I) on short circuit, of an alternator that is running at a rated frequency.

**SCRAM** Static Column dynamic Random-Access Memory. A dynamic random-access memory (DRAM) organized as 64K × 1 using one transistor cell arrays and housed in a 16-pin

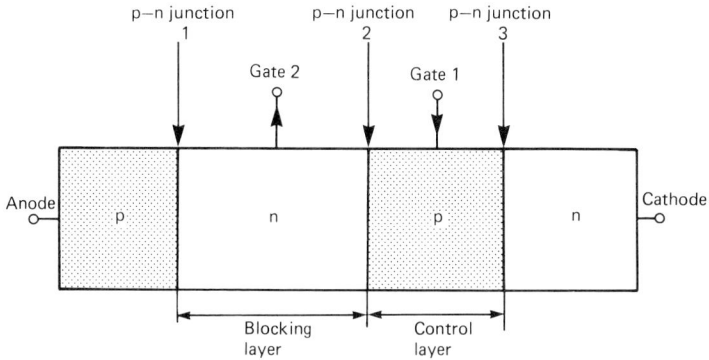

*Figure S.3* A basic silicon-controlled switch (SCS)

package. The memory on the same row address can be accessed only by selecting column address without the aid of a clock, as on a static random-access memory (SRAM). In this mode, cycle time can be equal to an access time while a DRAM usually requires a cycle time of twice the access time. Therefore, the access time becomes competitive to current high-speed SCRAM.

**SCS** Silicon-Controlled Switch. A four-layer device similar to a thyristor but having both gate leads brought out. A basic SCS is shown in *Figure S.3*. It can be switched on by either a positive pulse to its cathode gate or a negative pulse to its anode.

**SDA** (1) Source Data Acquisition. Information being entered directly into a computer at the location where the data originates. An example of this is a point-of-sale (POS) system. (2) Source Data Automation. The coded recording of information on tags, paper tape (PT) or punch cards, which can be used repeatedly without rewriting being necessary in order to produce other records of data.

**SDF** Supergroup Distribution Frame. Used in a frequency-division multiplexing (FDM) system, this frame provides for the interconnection of equipment and apparatus which transmit signals having frequencies in the 60-channel basic supergroup range.

**SDLC** Synchronous Data Link Control. A uniform discipline employing synchronous data techniques to transfer data between stations in a multipoint, point-to-point or loop configuration.

**SDM** (1) Space-Division Multiplexing. The combination of a number of separate wires or fibres into a bundle or cable, so that each fibre or bundle can be used as an independent communications path, channel (CHNL(d)) or set of channels. Typically a multiplexing arrangement might employ time division multiplexing (TDM) on each SDM pair in an optical cable. *See* OSDM, optical space-division multiplexing.

(2) Selective Dissemination on Microfiche. In response to subscriber's indicated areas of interest, this service provides them with microfiche copies of documents. *See* COM, computer output microfilm (COM).

**SDR** Statistical Data Recorder. A facility that logs the cumulative errors status of an input/output (I/O) device of a disk-operating system (DOS) on the system recorder file.

**Se** Selinium. A semiconductor element of atomic number 34. Its grey allotrope form is very light-sensitive and is used extensively in photoconductive photocells.

**SE** (1) Sign Extend. Used while multiplying or dividing during a shift, this operation ensures that the binary digit (BIT) shifted is identical to the sign binary digit (DGT).

(2) Stop Element. The last element in each character used in start-stop transmission that has a minimum duration assigned to it. During the transmission the receiving equipment is placed at rest to prepare for the reception of the next character.

**SECAM** Système En Couleurs À Mémoire. A line-sequential colour television (TV) system originating in France and used in several other countries including the USSR. In operation the signal which conveys information about the luminance of the television picture is transmitted by amplitude modulation (AM) of the vision carrier. Information concerning colour is transmitted by frequency modulation (FM) of the vision subcarrier. Colour saturation and hue (chrominance) signals are sent separately on alternate lines.

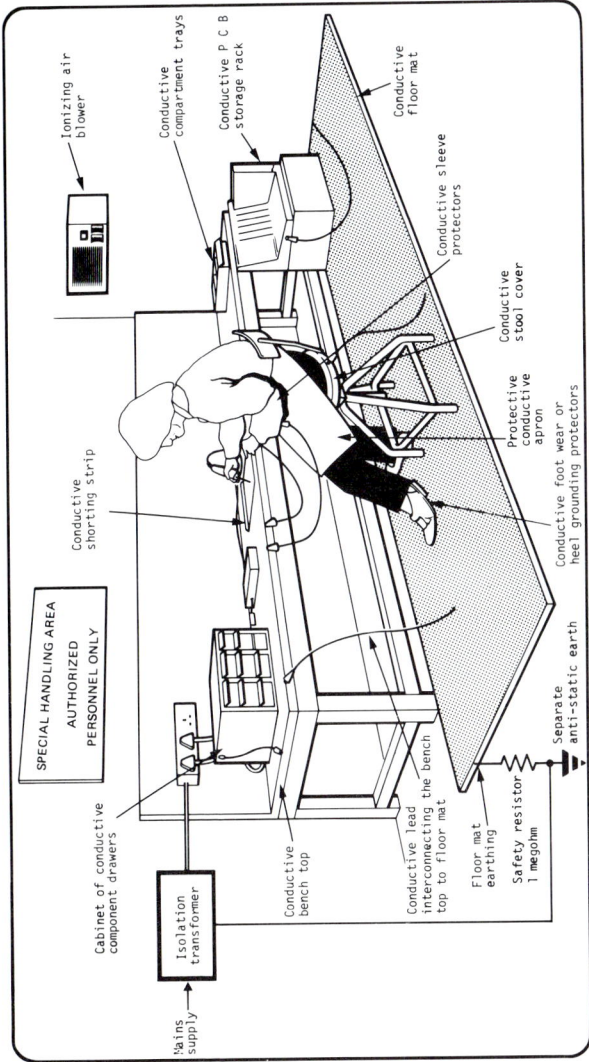

*Figure S.4*  The general layout of a special handling area (SHA)

**SELFOC**  SELf-FOCusing optical. A frequency selective optical fibre (OF) that can focus its output (O,2) transmission wave by controlled geometry, refractive indices and other factors.

**SEM**  Scanning Electron Microscope. An electron microscope with a beam that can scan an object field in a controlled or regular manner. It has many industrial uses; for instance it can be employed to expose resist material in high-resolution scanning patterns in the fabrication of integrated optical circuit (IOC) components.

**SEMCOR**  SEMantic CORrelation. A system of indexing that is computer aided.

**SERC**  Science and Engineering Research Council. Formerly the Science Research Council (SRC), this UK organization funds a number of research programmes into various aspects of science, engineering and computing.

**SERT**  Society of Electronic and Radio Technicians. A professional UK institute that provides services for technicians and technician engineers in all branches of electronics.

**SF**  Skip Flag. Typically a 1-binary digit (BIT) memory location in a microcomputer that represents a true or false skip or no-operation (NOP) condition, with respect to the instruction being executed by the processor.

**SFCS**  Slats and Flaps Control System. An intelligent avionics system, used for controlling the powerful slats and flaps of an airliner with safety. Its design incorporates two different types of 8-bit microprocessor (MP) independently programmed to ensure that the equipment is fail/safe.

**S/H**  Sample and Hold. An analogue circuit that is capable of capturing and retaining a signal, so that it may be converted by an analogue-to-digital converter (ADC).

**SHA**  Special Handling Area. An area for handling static sensitive devices (SSD) that is essentially free of electrostatic charge. The environment in an SHA is controlled in such a manner as to cause any electrostatic discharge to decay. This is achieved by the use of conductive materials for furnishings, tools and fittings, which can be augmented by the passing of ionized air across the working area. The general layout of an SHA is shown in *Figure S.4.*

**SHF**  Super High Frequency. A frequency between 3 and 30 gigahertz (GHz) forming part of a larger continuous range of frequencies. SHFs are considered by international agreement to be radio frequencies in band number 10 and have wavelengths between 1 and 10 cm.

**SHORT**  SHORT circuit. An abnormally relatively low-resistance connection between two points on a circuit (CCT).

Such a condition may be accidental or unintentionally caused, for example when two wires touch that under normal condition are insulated or separated from each other.

**SI** (1) Système International d'Unités (International System of Units). A coherent system of units based on the metre, kilogram, second, coulomb and dependent units on which international agreement has been reached by the General Conference on Weights and Measures. The three classes of units are *base units, derived units* and *supplementary units.* There are seven base units: the metre (m), the kilogram (kg), the second (s), the ampere (A), the kelvin (K), the candela (cd), and the mole (MOL). Derived units can be formed by combining base units. Thus, the unit of force (the newton, N) can be produced by combining the first three base units. Two supplementary units at present defined are the radian and the steradian, which are the units for plane and solid angles respectively. Prefixes are shown in the table below.

**SI Prefixes and their related Multiplication Factors and Symbols**

| Multiplication factor | | Prefix | Symbol |
|---|---|---|---|
| 1 000 000 000 000 000 000 | $= 10^{18}$ | exa | E |
| 1 000 000 000 000 000 | $= 10^{15}$ | peta | P |
| 1 000 000 000 000 | $= 10^{12}$ | tera | T |
| 1 000 000 000 | $= 10^{9}$ | giga | G |
| 1 000 000 | $= 10^{6}$ | mega | M |
| 1 000 | $= 10^{3}$ | kilo | k |
| 100 | $= 10^{2}$ | hecto | h |
| 10 | $= 10^{1}$ | deca | da |
| 0.1 | $= 10^{-1}$ | deci | d |
| 0.01 | $= 10^{-2}$ | centi | c |
| 0.001 | $= 10^{-3}$ | milli | m |
| 0.000 001 | $= 10^{-6}$ | micro | $\mu$ |
| 0.000 000 001 | $= 10^{-9}$ | nano | n |
| 0.000 000 000 001 | $= 10^{-12}$ | pico | p |
| 0.000 000 000 000 001 | $= 10^{-15}$ | femto | f |
| 0.000 000 000 000 000 001 | $= 10^{-18}$ | atto | a |

(2) SuperImpose. A process used in computer operations where data are moved from one location to another. Characters or binary digits (BITs) are superimposed onto the contents of a specified location.

(3) Serial Input. A method of data transfer between a computer and a peripheral device where information is transmitted for input to the computer, binary digit (BIT) by binary digit over a single circuit.

199

(4) Sample Interval. The discrete period of time (regular or irregular) over which the process of obtaining the value of a variable is made.

(5) Shift-In character. A control character that is able to bring a standard character set back into operation. It reverses the effect of the shift-out character (SO).

**SID** Sudden Ionospheric Disturbance. Unexpected radio fade-out caused by sudden short duration and abnormal ionization increases in the ionosphere D and E layers.

**SIGLE** System for Information on Grey Literature in Europe. A system for detecting, identifying and collecting widely available although unpublished literature. It is sponsored by the Commission of European Communities.

**SIL** (1) Single In-Line. A concept that provides standardized location and connection of components and circuits, on the basis of a row of connections on a 0.1 inch pitch. The term can also be applied to the pins on a single in-line package (SIP) or to the holes in a socket or printed circuit board (PCB).

(2) Systems Implementation Language. A term that can be applied to a number of different high-level computer programming languages. These can either produce very efficient code or have other particular features that enable them to be used for writing systems programs.

**SIMD** Single Instruction stream, Multiple Data stream. A particular type of design and construction of a computer. It can include in its make-up vector machines, parallel processors having one control unit and a number of arithmetic and logical units (ALUs), as well as some associative processors.

**SiO$_2$** Silica. An abundant natural compound that occurs in a number of forms including quartz and sand. It is an important source of the silicon that is used for the manufacture of electronic devices, components, integrated circuits (ICs) and as natural quartz for its piezo-electric properties.

**SIO** Serial Input/Output. A method of data transfer between a computer and a peripheral device. The data are transmitted for input to the computer or out to a peripheral device, binary digit (BIT) by binary digit over a single circuit.

**SIP** Single In-line Package. A package to house networks, components or circuits, with one row of connections which extend down at 0.1 inch intervals.

**SIPO** Serial In Parallel Out. A complex structure of shift registers in a large-scale integrated (LSI) circuit which is created using either flip-flop (FF,2) elements for static operation or simple clocked (CLK) invertors without feedback.

**SIR** Selective Information Retrieval. The access of particular items of information held on computer files, in response to requests made by users indicating a particular area of interest.

**SIS**  Semiconductor-Insulator-Semiconductor. Typically a solar cell constructed in a sandwich form, consisting of a transparent conductor such as the oxides of tin, indium, zinc or cadmium for the top contact layer. The second layer is a very thin insulating layer and the bottom layer is a p-doped (p-type) semiconductor. The top conducting oxides are actually heavily doped semiconductors; hence the resulting device is known as an SIS solar cell.

**SISD**  Single Instruction stream, Single Data stream. A type of design and construction of a computer that is accepted as the normal or conventional make-up.

**SITA**  Société Internationale de Tèlècommunications Aeronautiques. An organization that operates a message and data network for a large group of airline companies.

**SL**  Single Lead. An SL cable is a three core (three conductor) cable that has each core sheathed with single layer of lead as a covering.

**SLAC**  Subscriber Line Audio processing Circuit. The base for analogue/digital (A/D) conversion to a subscriber's line in an electronic telephone exchange. *See* SLIC.

**SLANG**  Systems LANGuage. A computer programming language in which statements normally correspond directly with machine level instructions and conversely, each machine operation is matched in a high level language statement.

**SLD**  Super-Luminescent Diode. A light-emitting diode (LED) with a narrow spectral width and a high intensity of electromagnetic radiation. It is normally used as a source for optical fibre (OF) transmission systems. It exhibits both spontaneous light emission (as in a LED) and stimulated light emission (as in a LASER) simultaneously. Its output (O,2) has a narrow spectural width and a higher radiance than an ordinary LED.

**SLIC**  Subscriber Line Interface Circuit. This adapts the subscriber's line to the exchange and provides the telephone with current, drives both ring and test relays and supervises the line. *See* SLAC.

**SLIC/SLAC**  Subscriber Line Interface Circuit/Subscriber Line Audio processing Circuit. These contain all the line functions required by an electronic telephone exchange.

**SLSI**  Super-Large-Scale Integration. Solid state circuitry usually on a single chip of silicon that has typically 100 000 or more transistors per chip.

**SLT**  Solid Logic Technology. Technology for the production of microelectronic circuits known as logic circuits because they transmit and control electrical impulses representing information within computing devices. The tiny devices operate at speeds which are as fast as six thousand millionths of a second.

**SM** Surface Mount. A system that enables connections to be made on the surface of printed circuit boards (PCBs). It is ideal for stacking feed through connectors for PCB applications.

**SMART** System for the Mechanical Analysis and Retrieval of Text. Designed for the interactive reviewing or searching of full text documents, this system uses particular words and phrases to search. It analyzes the text to produce a list of documents placing those incorporating the search words at the top.

**SMF** Systems Management Facilities. Available on various machines, this group of utilities makes possible the management of the particular system for tasks such as counting routines.

**SMPS** Switch Mode Power Supply. A power supply unit (PSU) that employs a switching technique. Its implementation provides considerable savings in size, weight and efficiency.

**SMT** Surface Mount Technology. The scientific and technical activities, research and development (R&D), applied towards the development of surface mount (SM) products, systems or services.

**SMX** Semi Micro Xerography. Producing xerographic copies of documents, this technique enables micrograph forms of documents to be input into a computer by means of a special reader.

**S/N** Signal-to-Noise ratio. *See* SNR.

**SNA** Systems Network Architecture. A particular design and construction for a distributed system, allowing computation of jobs at a number of physically separated locations. It also permits a number of devices to be interconnected in a variety of ways.

**SNAPSHOT** A colloquial expression for a dynamic dump of the contents of a particular memory storage location and register, which are being employed during a program execution.

**SNR** Signal-to-Noise Ratio. Usually expressed in decibels (dB), this is a comparison of the magnitude of a single to that of noise (unwanted disturbance) at the same point in a circuit. The form of its measurement and expression can vary depending upon the type of signal and noise concerned.

**SO** Shift-Out character. A character that is used to extend a standard character set by substituting it with another character set. Normally this is done to provide access to additional graphic characters.

**SOH** Start Of Header. A control character used in communications that is placed at the start of a sequence of characters comprizing a machine sensible address or routing information. The sequence of characters is referred to as a 'header'.

**SOLINET** SOuth-eastern LIbrary NETwork. A network connecting together over 200 libraries in the SE of the USA, thus permitting them to share facilities such as data processing (DP) and access to bibliographic information.

**SOM** (1) Start Of Message. A single character or group of characters used in telecommunications by a terminal to provide indication to other stations that the following transmitted information contains the addresses of those who should receive an answering message.

(2) Self-Organizing Machine. A machine that conducts its own internal organization, requiring no external intervention.

**SOP** Standard Operating Procedure. A particular mode of operation that is regularly or commonly used.

**SOS** Silicon-On-Sapphire. Materials in layers that produce devices capable of achieving bipolar speeds by using MOS technology, achieved by insulating the circuit components from each other. SOS can also refer to the process of fabricating these devices.

**SP** SPace character. (a) A character used in data processing and telecommunications. It is a non-printing character that is employed as a blank to separate words, etc. (b) An area in a memory that consists of blank characters. (c) A binary zero condition on a channel (CHNL(c)) or line during data communications.

**SPADE** Single-channel-per-carrier, Pulse-code-modulation, multiple-Access Demand-assignment Equipment. A system that has been developed for use in association with INTELSAT IV satellites to make more efficient use of their transmission capacity. It allows each carrier to be transmitted only when actually required for speech, enabling the transponder in the satellite to handle a much larger number of circuits before becoming overloaded.

**SPC** Stored Program Control. A program held in an electrically alterable memory or store that contains appropriate instructions to effect a method of control. SPC is often used to describe processor-controlled telephone switch systems that are controlled in the same manner, although differing widely in construction.

**SPDT** Single-Pole Double-Throw. A construction of switch that can be incorporated into a single line or pole. It has a single centre or common terminal with a movable blade. When activated (thrown) the blade is able to make contact, in one direction or the other, with one of two available terminals.

**SPEC** SPECification. An approved or accepted requirement or criterion for controlling technical performance or practice. It is by necessity precise, thoroughly developed and written down, updated to conform with current practices and used widely in order to be most effective.

**SPICE** Simulation Program with Integrated-Circuit Emphasis. A program used for the analysis of integrated circuits (ICs). It is a general purpose circuit simulator which can be used for three main analysis techniques: (i) linear alternating current (AC) analysis, (ii) non-linear direct current (DC) analysis, and (iii) non-linear transient analysis.

**SPIN** Searchable Physics Information Notes. A store of publications of the American Institute of Physics held on computer files.

**SPL** (1) System Programming Language. Developed for writing systems orientated software packages, this computer language's syntax and sematics are used to produce small and fast loading modules for specific equipment.

(2) SPLice. The action of joining two or more cables together providing suitable protection for the joint or joints such that the final assembly is as close as practical to a continuous cable. It is also used to describe a completed joint.

**SPRAT** Small Portable RAdar Torch. A lightweight, about 2.5 kilogram (kg), hand-held radar system that is based on a Gunn diode and has an effective range of around 600 metres (m,2).

**SPS** (1) Symbolic Programming Systems. A computer programming language that represents terms by quantities and locations.

(2) String Process System. A package of computer software that has been designed and assembled for the manipulation and processing of consecutive sets or strings of characters.

**SPST** Single-Pole Single-Throw. A construction of switch that can be incorporated into a single line or pole. It has a single terminal with a moveable blade which is able, when activated (thrown), to make contact with another single terminal.

**SPX** SimPleX. Pertaining to a circuit or channel that permits operation or transmission in one direction only, as opposed to duplex (DX).

**SQA** System Queue Area. Maintained by an operating system (OS), this is a main memory storage area specifically for control, status information or tables.

**SQUID** Superconducting Quantum Interference Device. A device produced by using conventional thin-film techniques, from superconductors such as Niobium, *see Figure S.5*. In its

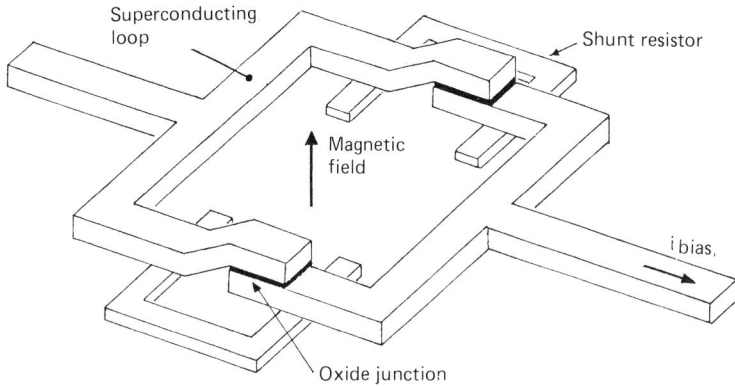

*Figure S.5* An example of a planar thin-film superconducting quantum interference device (SQUID)

analogue mode it will detect magnetic fields down to as low as $10^{-17}$ tesla. In its digital applications, it combines unprecedented high switching speeds with low dissipation for 100 GHz binary dividers, as well as uses in very-large-scale integration (VLSI) gate and memory arrays for a superconducting computer which is currently being investigated by International Business Machines (IBM).

**SRAM** Static Random-Access Memory. A random-access read/write memory (RAM) that is able to retain data indefinitely so long as it has an adequate power supply. An SRAM can be implemented in either bipolar or metal-oxide semiconductor (MOS) and is basically a read/write memory that stores information or data in integrated flip-flops (FF).

**SRBP** Synthetic Resin Bonded Paper. A laminated sheet of insulating material consisting of paper bonded together with synthetic resin under the action of pressure and heat. It is available in a number of grades in thicknesses from 0.2 to 50 mm, and is widely used for circuit boards, panels and terminal boards.

**SRC** Science Research Council. Now known as the Science and Engineering Research Council (SERC), this organization funds a number of research programmes into various aspects of science, engineering and computing.

**SRCNET** Science and engineering Research Council NETwork. A packet-switched data transmission network in the UK run by the Science and Engineering Research Council (SERC).

**SRIM** Selected Research In Microfiche. An information service held on microfiche that can provide document reference concerning new material.

**SRL**  Structural Return Loss. A measure of the reduction in power (loss) suffered by coaxial cables due to various irregularities in their structural geometry which in turn gives rise to forward and backward echoes.

**SSB**  Single SideBand. A transmission method where either of the frequencies above or below the carrier of an amplitude modulated (AM) wave is used, the unwanted frequency (or sideband) as well as the carrier being filtered out.

**SSC**  Station Selection Code. A code employed to indicate to a computer that access from a specific terminal or data processing (DP) station is required by the user.

**SSD**  Static Sensitive Device. An electrical or electronic device that can be irreparably damaged as a result of electrostatic charges, usually via its terminals. SSDs can include metal-oxide semiconductor (MOS) devices, junction field effect transistors (JFETs), beam lead diodes, hybrid integrated circuits (ICs), panel electronic circuits (PECs), some bipolar transistors and other small junction area devices.

**SSI**  Small-Scale Integration. The design and production of integrated circuits (ICs) or microcircuit chips (CHIPs). The solid state circuitry is usually produced on a single chip of silicon having fairly simple circuits with fewer components in a given area than medium-scale integration (MSI). Typically an SSI circuit contains from one to four logic circuits.

**SSR**  (1) Solid State Relay. A relay having all its constituent parts made from solid state components and involving no mechanical movements. The lack of physical contacts and no moving elements increases the mean time between failure (MTBF), decreases problems such as radio frequence interference (RFI), provides resistance to corrosion, enables usage in explosive environments and provides rapid switching speeds.
(2) Secondary Surveillance Radar. An essential facility for aircraft identification and control normally employed as an integral part of air traffic control systems. It is a co-operative system in which civil aircraft are fitted with a transponder, which is able to reply with a coded signal to an interrogation signal made by the SSR.

**SST**  Single-Sideband Transmission. An amplitude modulation (AM) transmission system in which only one of the sidebands (frequencies above and below the amplitude modulation carrier wave) is transmitted, the unwanted carrier and sideband being suppressed.

**stat-**  A prefix of a unit used in the now obsolete centimetre, gram and second (CGS) electrostatic system of units.

1 stat ampere = $3.336 \times 10^{-10}$ ampere (A)

1 stat volt = $2.998 \times 10^2$ volts (V)

1 stat ohm = $8.988 \times 10^{11}$ ohms

**STD** (1) Subscriber Trunk Dialling. A form of service provided by a telephone exchange in which subscribers are able to call other subscribers outside their own local area without operator assistance. All the UK is now on STD. It is also known as 'direct distance dialling' (DDD).

(2) STanDard. An approved or accepted requirement or criterion for controlling technical performance or practice. It is by necessity precise, thoroughly developed and written down, updated to conform with current practices and used widely in order to be most effective.

**STDM** Synchronous Time Division Multiplexor. Able to share a synchronous data link, an STDM is a multiplexor (MUX) that scans and interleaves binary digits (BITs) into the frames of the incoming data.

**STI** Scientific and Technical Information. A generic term used to describe data of a scientific or technical nature.

**STL** Schottky Transistor Logic. A form of integrated injection logic ($I^2L$) where the collectors are formed as Schottky barriers to reduce any excessive storage of charge at the collector junction. This results in an increased speed of operation as well as a reduction in the total voltage swing.

**STP** Standard Temperature and Pressure. Used in physical measurement these reference quantities or values are 0° centigrade (C) for temperature and 760 millimetres (mm) of mercury or 101 325 pascal (Pa) for pressure. Also known as 'normal temperature and pressure'.

**STRESS** STRuctural Engineering System Solver. A computer language employed in civil engineering for finding solutions to structural analysis problems.

**STRUDL** STRUctural Design Language. A computer language, developed from the structural engineering system solver (STRESS) language, which is employed in the analysis and design of structures.

**STTL** Schottky Transistor-Transistor Logic. A transistor-transistor logic (TTL) device that employs Schottky diodes for an improved speed of performance.

**STX** Start of TeXt. A control character used in communications that is placed at the head of a sequence of characters (text) and is transmitted through to the ultimate destination. A heading sequence started by a start of header (SOH) can be finished by an STX character.

**SUB**  SUBstitute. (a) A specific control character that is used to replace an invalid character. (b) The action of replacing a particular element of information with another element of information.

**SUHL**  Sylvania Universal High-speed Logic. Introduced in the early 1960s by Sylvania in opposition to Texas Instruments' (TI) transistor-transistor logic (TTL) range. At the time there was considerable rivalry between the merits of the two series until the market decided in favour of TI.

**SVC**  SuperVisor Call. An instruction that interrupts (INT) the program which is executing it, and switches control to an executive routine.

**SVR**  Super Video Recorder. Grundig's particular format used for video cassettes and recorders.

**SWAP**  Stewart-Warner Array Program. A chip (CHIP) of transistors and resistors manufactured with connections implemented in a user specified manner.

**SWG**  British Standard Wire Gauge. A legal UK system that employs numbers to designate the diameters of wires.

**SWOP AMP**  SWitchable input OPerational AMPlifier. A precision operational amplifier with two identical input stages sharing a common load and output stage, as shown in *Figure S.6*. One input at a time is selected using transistor-transistor logic (TTL), complementary metal-oxide semiconductor (CMOS) logic or user programmable logic levels. The

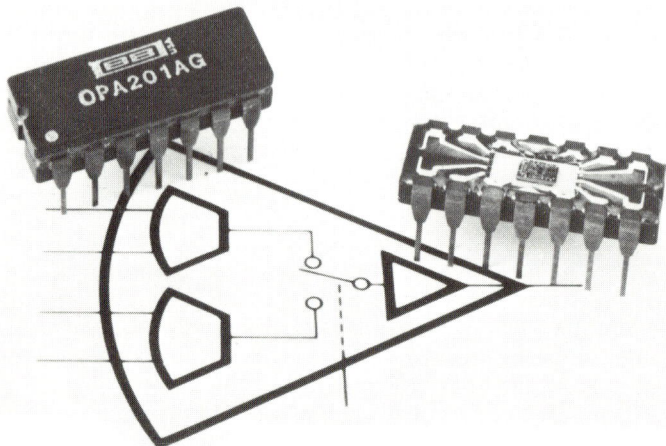

*Figure S.6* A switchable input operational amplifier (SWOP AMP); this particular equipment has two inputs and a single output (*courtesy* Burr-Brown International Ltd)

unselected input circuitry exhibits a very high input impedance drawing less than one picoamp. Applications include extremely-low drift amplifiers, multiplexed input amplifiers with good matching, switchable gain circuits and auto zero systems.

**SWR** Standard Wave Ratio. The ratio of the magnitude of an anti-node compared to a node in a field pattern of a transmission line produced by two similar frequency waves propagating in different directions.

**SYN** (1) SYNchronization signal unit. Used in common channel signalling (CCS), this group of information and check binary digits (BITs) contains distinctive information that can swiftly set up synchronism between sending and receiving terminals in a signalling network.

(2) SYNchronous idle character. A control character, used in synchronous data transmission, that is transmitted to set up synchronism between one data terminal and another. It can also maintain synchronism when no other character is being transmitted.

(3) SYNchronizing pulses. Pulses that are employed in the transmission of television (TV) pictures to provide a stable reference frame.

**SYSGEN** SYStem GENeration. (a) Used in some computer systems to create a supervisor and delay management support for a particular system configuration. Also used to include the program products that have been ordered. (b) The action of producing an operating system (OS) by selection from a large number of modules. The units are modified to create the optimum system.

**SYSLOG** SYStem LOG. A data set that is capable of storing job-related information, operational data, descriptions of unused occurrences, commands and messages to or from an operator.

**SYSTRAN** SYStem TRANsatlantic. An automatic system of machine translation (MT,3) that provides translations between English, French, Russian and Spanish.

**SXS** Step-By-Step switch. A switch that moves in such a way that it is synchronized with a pulse device such as a rotary telephone dial. Every digit that is dialled causes the movement of successive selector switches which carry the connection forward until the desired telephone line is reached. It is also known as a 'stepper switch'.

**T**

**T** (1) T antenna. Comprizing a horizontal conducting element insulated at its ends with a down lead connected at its midpoint, this type of antenna may have an element made up from a number of conductors connected in parallel.

(2) T network. An unbalanced network circuit consisting of three impedances, two series elements and a shunt element connected between their junction and earth, as shown in *Figure T.1*. It is also known as a star or Y- network.

(3) A flip-flop (FF) circuit that has one input (sometimes two); when a pulse appears at the input(s) this will cause the circuit to change its state to one of two possibilities.

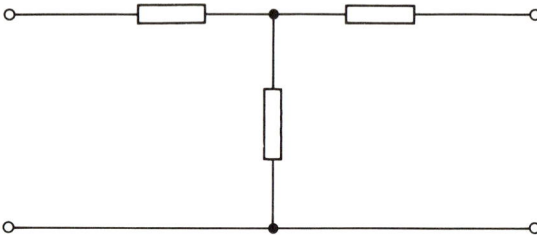

*Figure T.1*   A T-network

**Ta**   TAntalum. A metal of atomic number 73, having extremely high resistance to corrosion.

**TAB**   (1) TABulation character. A control character that is employed to move a printing mechanism or cursor in either a vertical or horizontal direction.

(2) Tape-Automated Bonding. A system used to mount semiconductor chips (CHIP) onto flexible continuous printed circuits, which speeds production and improves reliability. All the inner leads are brought into contact simultaneously with the bonding pads on the chip. *See Figure T.2*. The substrate comprizes a continuous strip of Kapton plastics film, with sprocket holes for automatic advance and precise alignment. It is also known as 'gang bonding'.

**TASI**   Time Assignment Speech Interpolation. In a telecommunication voice circuit audible speech is present for approximately 45% of the time. TASI switching equipment can connect a party to an idle circuit while speech is taking place and disconnects the party when speech stops, so that a different party can use the same circuit. The efficiency can be improved from 45% to approximately 80%.

**TAT**   TransATlantic cable. A submerged telecommunications cable. The designation of such cables are identified by digits, for example TAT-5.

**TBMT**   Transmitter Buffer eMpTy. One of the five status binary digits (BITs) of a standard universal asynchronous receiver/transmitter (UART).

**Tbps**  TeraBits Per Second. A signalling rate of $10^{12}$ binary digits (BITs) per second, that is one million bits per second.

**TC**  (1) ThermoCouple. The symbol TC is used to represent thermocouple on parts lists and to annotate schematic diagrams.

(2) Toll Centre. A class 4 office where telephone operators are present; if they are not, then it is called a toll point (TP). It is one of five class numbers assigned to offices according to their function in the US direct distance dialling (DDD) telephone network. Any one centre can handle telephone call traffic from one, two or more centres lower in the hierarchy, class 1 being the top classification.

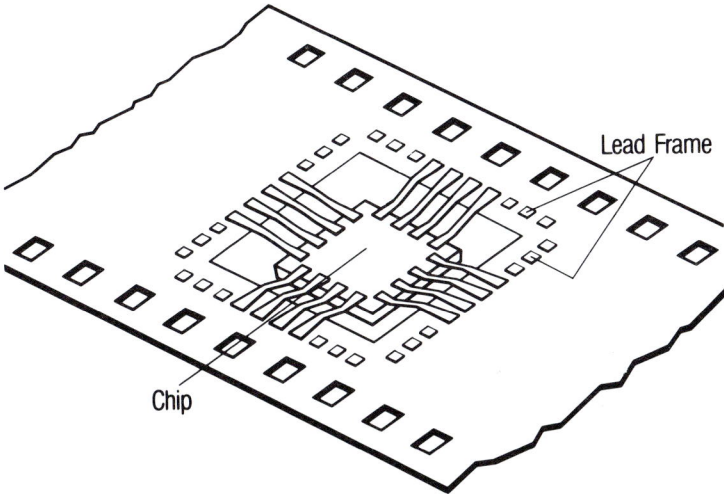

*Figure T.2*  The tape-automated bonding (TAB,2) process bonds leads simultaneously (*courtesy* Du Pont de Nemours International SA)

**TCAM**  TeleCommunications Access Method. A subsystem that controls the transfer of messages between an application program and remote terminals, providing a high-level message control language. It can be tailor made to customer application requirements.

**TCB**  Task Control Block. A memory location containing information or a set of routines used to control or manage jobs or tasks within a computer system.

**TCBM**  Time-Consistent Busy Hour. Beginning at the same time on each week day, a period during which the highest average

211

telecommunication traffic is measured. A TCBM is an uninterrupted period of 60 minutes.

**TCM**  Terminal to Computer Multiplexor. A unit that is able to control the use of several sending and receiving channels (CHNL(b)) between a terminal and a computer simultaneously.

**TCP**  (1) Transmission Control Program. A program within a computer network that sets into motion formalized methods of communication between networks.

(2) Tape Conversion Program. A special computer program used for copying and/or converting paper tapes (PT) from one tape format to another, e.g. from hexadecimal (HEX) to binary format or from binary to hexadecimal format.

**TCS**  Terminal Control System. Designed to manage multiterminal operations in a computer system, the TCS schedules the usage of hardware and all input/output (I/O) processing.

**TCU**  Transmission Control Unit. An input/output (I/O) control unit that provides address information for messages to and from a number of remote terminals; it also acts as an interface between a central processor unit (CPU) and the terminals.

**TD**  Transmitter-Distributor. A specific device in a teletypewriter (TTY) terminal that makes and breaks the line in timed sequence. The term is generally used when making reference to paper tape (PT) transmitters.

**TDM**  (1) Time Division Multiplex. A process, circuit (CCT) or device that enables the transmission of two or more signals over a common path by using successive time intervals for different signals.

(2) Time Division Multiplexing. A method of using a single path to provide a number of channels (CHNL(a)) by time-dividing the path into a number of time slots and then assigning each channel its own intermittently repeated time slot. At the receiving end, each time-separated channel is reassembled. Ideally suited for the transmission of digital (D,4) data the system is now used for digitized speech and other signals. Allocations of time-slots may be regularly repeated (*fixed cycle*), or may be made on demand (*dynamic*).

**TDMA**  Time Division Multiple Access. A technique that allocates different time slots to various users and enables communicating devices at separate geographical locations to share a multi-point or broadcast channel (CHNL(a)).

**TDR**  (1) Time Domain Reflectory. A system for making diagnostic measurements of fibre optics (FO) of new designs. It can detect and analyze variations of dispersion or dimensions along the fibre length. It is also known as a 'photoprobe' or 'radar'.

(2) Transmit Data Register. Employed in asynchronous communications interface adaptor (ACIA) transmission systems, this memory location is used to hold data that are ready to be transmitted out.

**TEAM** Terminology, Evaluation and Acquisition Method. A machine-aided translation (MAT) system located in Germany that is able to offer facilities in eight major European languages.

**TeD** TElDec. A video-disk system from Telefunken and Decca.

**TELE-** A prefix that denotes distance; for example, television is vision transmitted over a certain distance; a teleprinter is a remote printer; telephone is sound transmission.

**TELECOMM** TELECOMMunication. Any process that allows information of any nature transmitted in a useable form (e.g. audible signals, visible signals, printed copy, fixed or moving pictures) to travel from a sender to one or more receivers. The transmission is achieved by means of any electromagnetic system (e.g. electrical transmission by wire, radio, optical means and guided waves).

**TELEPUTER** A combined TELEvision and COMPUTER. Designed to be the central component in the home telecommunications services, this unit combines six major technologies; colour television, videotex, computing, video cassette recording, video disk and telecommunications. The system can be set to access automatically any number of computers, becuase it has a built-in personal computer capability. Data gathered can be manipulated, analyzed and processed at the terminal under the control of various access programs.

**TELEX** TELetypewriter EXchange service. An automatic dial-up dedicated switched network that provides for communication between teletypewriter (TTY) and teleprinter equipment worldwide. A terminal is used to produce information on a paper tape (PT) which is fed into a reader and then transmitted.

**TEMP** (1) TEMPerature. A measure of the average kinetic energy of the moving molecules or atoms in a material. Generally it is a measure of how hot or cold material is compared to a defined scale. The Système International de Unités (SI) unit of temperature is the kelvin (K), but the Celsius (C,8) or centigrade scale is normally used for everyday measurements.

(2) TEMPorary register. Typically a 12-binary digit (BIT) memory location that holds or latches an arithmetic and logical unit (ALU) operation, in order to avoid race conditions before it is sent to a final memory location.

**TERA-** A prefix denoting one million million, a billion ($10^{12}$); for example a terabit is $10^{12}$ binary digits (BITs).

**TFT** (1) Thin-film transistor. Fabricated using thin-film techniques, typical of a thin-film transistor, is an insulated-gate field-effect transistor (IGFET). It is manufactured on a substrate.

(2) Thin-Film Technology. Technology in which electronic circuits or elements are formed by a process in which vacuum-deposited or sputtered films are produced upon a supporting material. TFTs can be considered as films that are less than 5 micrometres thick as opposed to films greater than 10 micrometres thick.

**THERMISTOR** THERMally sensitive resISTOR. Usually fabricated from semiconductor material shaped as rod, bead or disk and named accordingly, thermistors are available in two forms. Negative temperature coefficient (NTC) thermistors that have a resistance that decreases with increasing temperature, and positive temperature coefficient (PTC) thermistors that have a resistance that increases with increasing temperature. Both forms cover a range of temperature coefficient from around $-65\%$ $+70\%$ per °C.

**THP** Through (the) Hole Plating. Printed circuit boards (PCBs) having circuits on both sides which are connected electrically by the plating that constitutes the circuit tracks passing through holes in the board.

**Ti** TItanium. Resembling iron, this metallic element has an atomic number 22, atomic weight 47.90, specific gravity 4.5 and melting point of 1850°C.

**TIMS** Transmission Impairment Measuring Set. Normally a portable unit designed for in-the-field characterization of analogue circuits for data communications. The sets are usually compatible with International Telegraph and Telephone Consultative Committee (CCITT) standards and measure impulse noise, random circuit noise, level versus frequency distortion and signal-to-noise ratio (SNR).

**TIP** Terminal Interface Processor. A small computer connecting terminals to a packet-switching network.

**TIPTOP** Tape InPut/Tape OutPut. A method for introducing data to an input device or reading data from an output (O,2) device. The technique can include the use of plastics or metallic magnetic tape (MAGTAPE), perforated paper tape (PT) or fabric tape loops.

**TKO** TrunK Offering. An exchange facility permitting an operator to intrude into an established connection to offer one of the parties an incoming long-distance call.

**T²L** Transistor-Transistor Logic. *See* TTL.

**TLU** Table Look Up. The action of finding a function value that corresponds to a stated or implied argument, the information being held in a table of function values stored in a computer.

**TLX** TeLeX. *See* TELEX.

**TM** Tape Mark. A special character written on a tape to signify the physical end of recording. It can indicate the actual end of the tape or it can be used to subdivide the tape into sections.

**TMP** TeMPerature. *See* TEMP.

**TMS** Transmission Measuring Set. Equipment made up of a calibrated signal generator and a level measuring set, that is used to measure the insertion loss or gain of a transmission path.

**TNC** Threaded-Nut Coupling. A type of connector used for coaxial cable (COAX) radio frequency (RF) interconnection. It consists of a coupling that employs a screw-on interface.

**TNEP** Total Noise Equivalent Power. A measure of the minimum detectable signal that can be found amongst noise in detection and multiplication processes. It can be defined as the amount of light falling on a photodetector, required to produce an output (O,2) signal equal to the noise output. It is usually expressed in nanowatts.

**TNS** Transaction Network Service. A data exchange system used in US metropolitan areas that provides a basic communications service for short data messages, normally enquiry-response financial transactions.

**TO** A designation for a package normally used to house discrete transistors. The enclosure is most often constructed from metal although some plastics versions are available.

**TOD** Time Of Day. (a) An electronic clock (CLK) device that is able to measure and record time in hours, minutes and seconds over a 24-hour cycle. When called by a central processor unit (CPU) it is able to provide the time. (b) A notation made to indicate the particular time of day an event occurred.

**TOS** (1) Tape-Operating System. An operating system (OS) held on a tape or a number of tapes. Some systems have compilers and linkage editors on a system tape and use tape libraries for user programs an data files.

(2) Temporarily Out of Service. A designation given to a telephone subscriber's line when it is temporarily deliberately disconnected from equipment at the exchange.

**TP** Toll Point. A class 4 office where operators are not present; if they are then it is called a toll centre (TC). It is one of five class numbers assigned to offices according to their function in the US direct distance dialling (DDD) telephone network. Any one centre can handle telephone call traffic from one, two or more centres lower in the hierarchy, Class 1 being the top classification.

**TPA** Transient Program Area. An area in the memory map (MAP,2) of a microcomputer using a control program for microcomputer (CP/M) disk operating system (DOS). It allows the user to write programs to run so they can access areas of basic disk operating system (BDOS), basic input output system (BIOS) and console command processor (CCP) to use routines already present in the system.

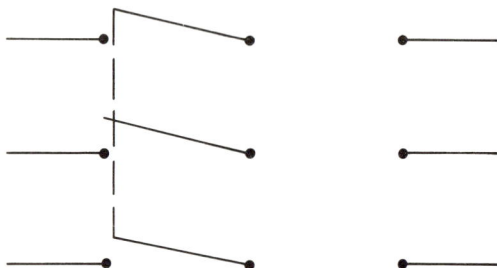

*Figure T.3* Symbol for a triple-pole single-throw (TPST) switch

**TPST** Triple-Pole Single-Throw. A switch that can be incorporated into a triple line or pole. It has three terminals each having a moveable blade. The blades are so constructed that when activated (thrown), each simultaneously makes contact with its respective line terminal. *See Figure T.3.*

**TR** (1) TRansistor. The symbol TR is used to represent transistors on parts lists and to annotate schematic diagrams.

(2) Transmit-Receive. A switch employed in a radar system, where the system has a common antenna for both transmission and reception. It decouples the receiver automatically from the antenna during a period of transmission. *See* ATR.

**TRAN** TRANsmit. The action of moving data or information from one location to another, normally via a transmission line or communications link.

**TRANSCEIVER** TRANSmitter/reCEIVER. A radio transmitter and receiver combined in a common enclosure employing some common circuit components for both transmitting and receiving, normally used for mobile or portable devices.

**TRANSPONDER** TRANSmitter/resPONDER. A device or circuit often used in aircraft and orbital satellites that is able to receive an input signal, act upon it and retransmit it. The device is used as a means of reaching a wide area with a radio transmitter.

216

**TRF**   Tuned Radio Frequency. An amplitude modulated (AM) receiver that has one or more parts of its radio frequency (RF) amplification before its detector, and audio-frequency stages after. It has been superseded by superheterodyne receivers.

**TRIAL**   Technique for Retrieving Information from Abstracts of Literature. Using the abbreviated representation of the contents of documents held on a computer file, this system is able to manipulate and process the information.

**TS**   (1) Time Sharing. The use of a computer by two or more users simultaneously in real-time (RT,1). Ultra-high-speed switching of processing time is employed to achieve this sharing.
(2) Time Shared. The action of computer systems undertaking time sharing (TS,1).

**TSC**   Transmitter Start Code. A code normally consisting of two letters that is used to call up a remote or outlying unit so that it automatically switches on its tape transmitter.

**TSO**   Time-Sharing Option. An option that is made available with some operating systems (OS), to provide interactive use by a number of operators.

**TSPS**   Traffic Service Position System. A stored program computer having telephone operator consoles that permit the efficient handling of calls requiring operator intervention.

**TSS**   (1) Time-Sharing System. An interactive computer operating system (OS) that has been designed so that the central processor unit (CPU) can be simultaneously time shared (TS,2) amongst the users.
(2) Time-Shared System. A computer system that is capable of being time shared (TS,2) but has not been specifically designed for this purpose. *See* TSS,1.

**TT**   TeleTypewriter. *See* TTY.

**TTL**   Transistor-Transistor Logic. One of the most commonly used logic families due to its versatility and high speed. It is a bipolar integrated circuit (IC) based on saturing transistors. The logic uses transistors with multiple emitters which are easy to fabricate, requiring only a single isolated collector region upon which a single base region and then several emitters are diffused. *See Figure T.4.* TTLs have high power consumption. Also known as $T^2L$.

**TTS**   TeleTypeSetting. A system of typesetting remotely where a typesetting machine is controlled by input from a punched paper tape (PT). This can be achieved over a communications link or directly.

**TTY**   TeleTYpewriter. An electromechanical unit comprising a keyboard and a printer. Originally used to transmit information

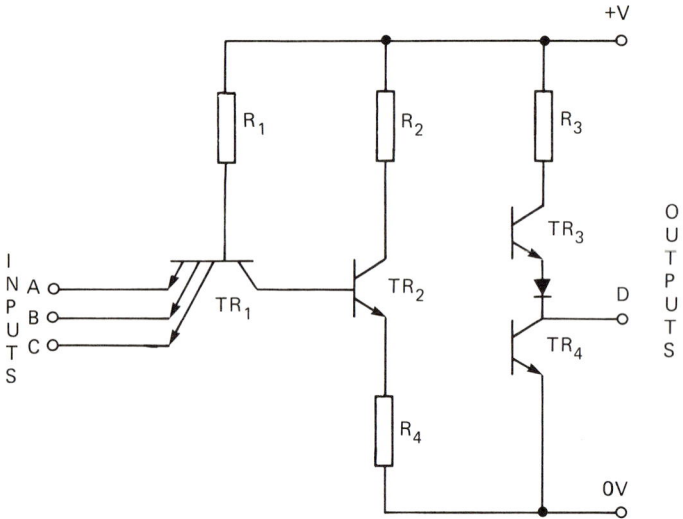

*Figure T.4*   A basic transistor-transistor logic (TTL) gate

in communication systems, it is now used to communicate with computers using American Standard Code for Information Interchange (ASCII) or other codes.

**TV**   TeleVision. (a) The technique of transmitting electrical signals in order to provide moving or stationary images with or without sound over a distance, via a suitable channel (CHNL(g)) or path. (b) A system that fulfils the techniques described in TV(a). (c) A receiver that is capable of receiving the signals produced by a system described in TV(b).

**TVI**   TeleVision Interference. Interference to a domestic television (TV(c)) set can be caused by a local radio transmitter, or be produced by unsuppressed electromechanical devices such as electric motors.

**TVT**   TeleVision Typewriter. A unit that is capable of displaying information on a television screen.

**TWEETER**   A high frequency loudspeaker designed to operate only on high audio frequencies in high-fidelity sound reproduction systems. It is usually part of a larger loudspeaker assembly that consists of two or more speakers. (*See* WOOFER).

**TWT**   Travelling Wave Tube. A hardware device that is employed to amplify microwave ($\mu$W) and other high frequency (HF) signals. Within the tube an electron beam interacts continuously with a radio frequency (RF) electromagnetic field

218

to produce amplification or, in some types of tube, oscillation at microwave frequency.

**TWX** TeletypeWriter eXchange service. A North American public switched teletypewriter (TTY) service that has suitably arranged teletypewriter stations where customers are able to rent a teletypewriter provided with lines to a central office for access to other such stations throughout the USA and Canada. The service uses both American Standard Code for Information Interchange (ASCII) and Baudot machines and equipment.

**TX** (1) Transmitter. A unit or circuit in a communication system that converts information (often audible) into electrical signals ready to be conveyed from one designated location to another by means of a transmission line, waveguide, radio channel (CHNL(a)) or wire. Normally on communications equipment a red panel indicator shows that a TX circuit is in operation. (2) TeleX. *See* TELEX.

## U

**U** Rack height. A 'U' value is equivalent to 1¾ inches (44.45 mm) when referring to the height increments of a 19-inch racking system; for example a 3U-high rack would be equal to 5¼ inches. The standard was first set by the American Post and is now a *de facto* standard IEC2g. The U values can also be equated to the PCBs cards or other items that fit into the racks.

**UA** User Area. Part of a computer disk where semipermanent data may be stored. It can also be used to store programs, subprograms and subroutines.

**UADS** User Attribute Data Set. A collection of information held on a computer file concerning the users of a particular computer system.

**UART** Universal Asynchronous Receiver Transmitter. A particular device that is used to interface a data terminal or word parallel controller to a communication network.

**UAX** Unit Automatic exchange. A UK designation given to an unattended automatic telephone exchange, that provides a service to a small community.

**UBC** Universal Block Channel. A direct memory access (DMA) channel (CHNL(b)) that operates in a block-mode, for high-performance peripheral controllers such as those used with disks or magnetic tapes (MAGTAPE).

**UBL** UnBLocking signal. A signal that is transmitted to a busy-out circuit (CCT) at its distant end. The UBL ends the busy signal, which may be continuous, indicating that the circuit is again available for use.

**UC** Upper Case. Capital letters of any type face. (*See* LC,1).

**UCLA** UnCommitted Logic Array. Consisting of partially processed wafers of standard chip (CHIP) design, the UCLA provides a compromise between the standard and custom device, as the customer is able to specify the interconnections that are necessary to provide the particular functions required.

**UCS** (1) Universal Character Set. A printing device with a UCS facility can allow a number of character sets to be used, thereby providing any standard typeface.

(2) User Control Store. A section of a memory circuit designed to hold the sequence of commands which determine operations. It is set aside for the user's microprogram.

**UDC** Universal Decimal Classification. An expansion of the Dewey decimal classification system (DDS,2). It is sometimes known as the Brussels system.

**UF** Ultra Filtration. The technology for molecular/colloidal separations. It employs a skinned membrane structure that has semipermeable properties.

**UHF** Ultra-High Frequency. A frequency occurring between 300 and 3000 MHz and forming part of a larger continuous range of frequencies. UHFs are considered by international agreement to be radio frequencies in band number 9 and have wavelengths of between 10 cm and 1 metre.

**UKITO** United Kingdom Information Technology Organization. A collection of UK companies that are concerned with the applications of information technology (IT).

**UL** Underwriters Laboratories. An organization in the USA that is primarily concerned with qualifying equipment to their rigorous safety standards.

**ULA** Uncommitted Logic Area. Consisting of partially processed wafers of standard chip (CHIP) design, the ULA provides a compromise between the standard and custom device, as the customer is able to specify the interconnections that are necessary to provide the particular functions required.

**ULSI** Ultra-Large-Scale Integration. A degree of magnitude greater than very-large-scale integration (VLSI). Where VLSI microprocessors (MP) contain tens of thousands of components, ULSI microprocessors contain in the region of 100 000 components.

**UMF** Ultra MicroFiche. Pages of text and graphics photographically reduced with such small images that 3000 pages can be mounted on a 4 × 6 inch film.

**UMI** Ultra MIcrofiche. *See* UMF.

**UNIX** Developed by Bell Laboratories, this particular collection of procedures for the operation of a computer is general purpose and has been designed to be multi-user.

**UPC**   Universal Product Code. A symbol comprising a series of parallel lines which represent a 10 number code, that has been adopted as standard by the food industry and can now be found on most food items. The first five numbers in the code indicate the item's manufacturer, the second five identify various features including size and brand, although the price is not included in the code. The code is read by an optical scanner which senses light reflected from it.

**UPS**   (1) Uninterrupted (or Uninterruptible) Power Supply. A system designed to provide continuous power (normally of computer quality) in all abnormal power situations. UPS problems can cover millisecond fast power dropout to long-term blackouts lasting hours or days but generally less than an hour's duration; rapid dips in voltage (V) to long-term lows; and quick spikes in power to long-term high-voltage levels.

(2) UPper Side band. The side band containing all the frequencies above that of the amplitude-modulated (AM) carrier.

**UPU**   Universal Postal Union. An agency of the United Nations that promotes international co-operation between the postal services of different nations.

**URL**   User Requirements Language. A high-level computer programming language that permits the user to employ a notation with which he is familiar.

**US**   Unit Separator. A control character employed to designate the logical segregation of or to indicate the boundaries of items of, data or information.

**USASCII**   United States of America Standard Code for Information Interchange. Usually known as ASCII (*see* ASCII).

**USASI**   United States of America Standards Institute. The former name of the American National Standards Institute (ANSI).

**USART**   Universal Synchronous/Asynchronous Receiver/ Transmitter. A peripheral device chip (CHIP) used for data communications. Programmed by the central processor unit (CPU) it can operate using almost any serial data technique. Accepting data characters from the central processor unit in parallel format, it converts them into a continuous serial data stream for transmission, whilst simultaneously being able to receive serial data streams and convert them into parallel data characters for the central processor unit.

**USITA**   United States Independent Telephone Association. Consisting of independent telephone companies from a number of countries, this US-based organization deals with regulations and technical standards.

**USRT** Universal Synchronous Receiver/Transmitter. A general purpose communications interface that allows a high-speed synchronous communications device to transmit data to, and receive data from, a microcomputer system. It connects to the device via separate lines.

**UTE** Union Technique de l'Électricité. The French national standards authority, that has a system of specifications operating in a way similar to the MIL specification (MIL,2).

**UUO** Unimplemented User Operation. An instruction on some computer systems issued by the user program to request the supervisor program to perform a function that the user program is not permitted to perform itself.

**UUT** Unit Under Test. Describes a functional printed circuit board (PCB) assembly that receives test or inspection. The designation UUT is often given to assemblies that are receiving analogue in-circuit and functional printed circuit board assembly test via automatic test equipment (ATE) systems.

**UV** Ultra-Violet. High-energy electromagnetic waves ranging in a wide band of wavelengths from 380 nanometres (which is just beyond the visible blue-violet) to about 10 nanometres. Radiation from the sun contains ultra-violet rays which ionize the upper atmosphere of the earth producing the ionosphere.

**UVCB** Under-Voltage Circuit-Breaker. A circuit-breaker that comes into operation when the supply voltage (V,2) in a particular circuit (CCT) drops below a predetermined value.

**UV ERASER** Ultra-Violet ERASER. A device, often in the form of a bench top cabinet, that provides a concentrated source of ultra-violet (UV) light. It is used to erase the information stored semi-permanently within UV sensitive devices such as ultra-violet erasable programmable read-only memories (UV PROM).

**UVP** Under-Voltage Protection. The action of defending a circuit (CCT) from voltage (V,2) lower than predetermined requirements and often associated with some form of standby power source. *See* UVCB, under-voltage circuit-breaker.

**UV PROM** UltraViolet erasable Programmable Read-Only Memory. A programmable read-only memory (PROM) that permits information to be stored semi-permanently, erased by concentrated shortwave ultra-violet (UV) and then reprogrammed with new information. The device is housed in a dual-in-line package (DIP) with a quartz top that is transparent to shortwave UV light.

**V** (1) Volt. Le Système International d'Unités (SI) unit of potential difference (PD), electromotive force (EMF) and electrical potential. It can be defined as the difference of potential between two points on a conductor when the current flowing is one ampere (A) and the power dissipated between the points is one watt (W).

(2) Voltage. The potential difference (PD) between two points in a circuit (CCT) or device.

**VAB** Voice Answer Back. An audio response unit that provides voice responses, from a pre-recorded digitally coded voice or disk-storage vocabulary, to telephone terminal enquiries. It can link a computer system to a telephone network.

**VAC** (1) VACuum. A perfect vacuum is a space from which gas has been removed and the pressure is zero. High-vacuum electron tubess achieve pressures of $10^{-6}$ mm Hg (1 mm Hg = 133.32 N/m$^2$).

(2) Voltage, Alternating Current. The potential difference (PD) between two points in a circuit or device, with a current (I) that has its direction in that circuit or device reversed with a frequency (*see* Hz, hertz) that is independent of circuit constants, i.e. an alternating current (AC).

**VACC** Value-Added Common Carrier. An organization that provides and sells the services of a value added network. These networks are built using traditional common carriers, connected to computers which permit new types of telecommunication tariffs to be offered. The network can be packet switched or message switched. The services offered include transmission of data charged for by the packet and the transmission of facsimile (FAX) documents.

**VAD** VApour phase epitaxial Deposition. The action of depositing materials during a vapour phase oxidation process. *See* VPE, vapour phase epitaxial.

**VAN** Value-Added Network. *See* VACC, value-added common carrier.

**VARACTOR** VARiable reACTOR. A semiconductor diode (D,5) that behaves as a voltage dependent capacitor (C,6), when operated with reverse bias. The device can be used in low power applications, for instance varactor tuning. High power versions are employed as almost lossless frequency multipliers. For very high frequency (VHF) applications, gallium arsenide (GaAs) is used, but generally silicon is the material employed.

**VARISTOR** VARiable resISTOR. Usually of a semiconductor type, these resistors have a symmetrical current (I) voltage (V) characteristic which can be formed by connecting two p-n junction (p-n) diodes (D,5) in parallel. This arrangement is very non-linear as its resistance varies greatly with voltage (V).

**VARMETER** Volt-AmpeRes METER. Calibrated in reactive volt-amperes, this instrument measures the reactive component of volt-amperes in a circuit (CCT).

**V band** Microwave ($\mu$W) frequencies occurring between 46 and 56 gigahertz (GHz) and forming part of a larger continuous range of frequencies covering 0.225 to 100 gigahertz, which are usually subdivided into bands designated by letters. V band frequencies have wavelengths of between 0.652 and 0.536 cm.

**VCC** Video Compact Cassette. Developed by Phillips, this is a particular system for recording television programmes and other video information onto magnetic tape (MAGTAPE) housed in a plastics enclosure.

**VCO** Voltage Controlled Oscillator. Having a frequency controlled by the adjustment of an input voltage (V), this oscillator forms part of a phase-locked loop (PLL) system.

**VCR** Video Cassette Recorder. A domestic videotape recorder using ½ inch tape cassettes. There are three similar although not necessarily compatible systems: Betamax, V2000 and Video Home Systems (VHS).

**VDC** Voltage, Direct Current. The potential difference (PD) between two points in a circuit or device with a current (I) that has a unidirectional flow in that circuit or device with a substantially constant magnitude and direction, i.e. a direct current (DC).

**VDE** Verband Deutscher Electrotechniker. A West German certified test laboratory that qualifies or approves equipment and components for commercial and home use throughout most of Europe. The agency qualifies equipment to both radio frequency (RF) emission and safety standards.

**VDI** (1) Visual Display Input. The information received by a visual display unit (VDU).

(2) Video Display Input. The information received by a unit used to display (D,6) video information, that is a device equipped with a cathode ray tube (CRT), normally a television (TV(c)) receiver or a monitor.

**VDL** Vienna Definition Language. A computer language used to define the semantics of other programming languages.

**VDM** Visual Display Module. A liquid crystal display (LCD) in the form of a matrix, supplied in a front panel of a unit which is tailored to suit Eurocard rack mounting equipment ranges.

**VDR** Voltage Dependent Resistor. Having a resistance that decreases greatly with an applied voltage (V) increase, these non-linear resistors are often constructed of silicon carbide pressed with a ceramic binder, formed into disks or rods and fired at 1200°C.

**VDT** Visual Display Terminal. Equipment with a cathode-ray tube for the visual display (D,6) of information, normally connected to a keyboard (KB) that is used for inputting information.

**VDU** Visual Display Unit. A device that provides a primary output method for alphanumeric or graphical information visually on the face of a cathode-ray tube (CRT), under computer program control. Data may be input by a user via a keyboard (KB,1) and/or some other manual method such as a light pen, cursor controls or function buttons.

**VF** Voice Frequency. Frequencies used for the commercial transmission of speech are generally accepted to lie within the range of 200 to 3000 hertz (Hz), although a VF may be described as any frequency lying within the human audible range.

**VF band** Voice Frequency BAND. A frequency range between 200 and 3000 hertz (Hz) that can provide quite intelligible and adequate speech transmission.

**VFO** Variable Frequency Oscillator. An oscillator that has an output (O,2) frequency capable of being varied quite considerably.

**VGR** Video Graphics Recorder. Often produced as a desk-top unit, a VGR provides hard copies of video displays. It has wide applications across computer graphics, engineering, drafting, design analysis, medical and industrial imaging.

**VHD** Video High Density. A video-disk system developed by JVC that employs a laser to record signals on a disk. It is a capacitive system in which the pick-up is guided along the appropriate track by special signals.

**VHF** Very High Frequency. Frequencies occurring between 30 and 300 MHz and forming part of a larger continuous range of frequencies. VHFs are considered by international agreement to be radio frequencies in band number 8 and have wavelengths of between 10 and 1 metres.

**VHS** Video Home System. A widely used ½ inch video cassette format that was developed by the Matsushita Electronics Corporation.

**VISC** Video dISC. An early video-disk system that employed a mechanical pick-up.

**VLF** Very Low Frequency. Frequencies occurring between 3 and 30 kHz and forming part of a larger continuous range of frequencies. VLFs are considered by international agreement to be radio frequencies in band number 4 and have wavelengths of between 10 km and 100 km.

**VLP** Video Long Player. A video-disk system developed by Philips/MCA.

**VLSI** Very-Large-Scale Integration. The design and production of integrated circuit (IC) or microcircuit chips (CHIP). The solid state circuitry is produced on a single piece of material, normally silicon. VLSI describes circuits with a capability of between 16 kilobits (KB,2) and one megabit.

**VM** Virtual Memory. A technique that provides the virtual appearance of a larger main memory, by permitting the user access to a second disk, which is an extension of the main memory. By employing this method a larger memory is simulated than actually exists.

**VME** (1) Virtual-Memory Environment. A virtual-memory (VM) operating system (OS) that is available on some computers.

(2) Versa Module Europe. A bussing protocol for digital applications developed by Mostek, Motorola and Signetics. It defines the mechanical printed circuit board (PCB) dimensions, which are Eurocard based, the connectors, which are to DIN 41612, and the highways, protocol and enabling systems. It has been designed so that various manufacturers' equipment can be interfaced using a backplane system.

**VNL** Via Net Loss. The net losses of trunks in the long distance switched telephone network of North America. The trunk is said to be in a VNL condition when it is an intermediate trunk in a longer switched connection.

**VODAS** Voice-Operated Device Anti Sing. A system that permits transmission in only one direction at a time, to prevent unwanted self-oscillation (singing) in a two-way telephone circuit.

**VOGAD** Voice Operated Gain Adjusting Device. A device employed in some radio systems to remove fluctuation from input speech and to send it out at a constant level. No restoring device is needed at the receiving end. It is similar to a compandor.

**VOL** VOLume. The magnitude of the complex audio frequency signals in an audio frequency (AF) transmission system.

**VOM** Volt Ohm Milliameter. A test instrument for measuring voltage (V,2) resistance and current (I). It is usually portable and has an analogue meter as a readout.

**VOX** Voice Operated device. A circuit or device that automatically actuates a desired operation or function when in the presence of an audio signal. Often used to give automatic changeover on many transmit/receive units.

**VPE** Vapour Phase Epitaxial. A method of producing optical fibres (OF) of low loss, based on the principle of vapour phase oxidation. Volatile gaseous feed materials are delivered to a reaction zone in an epitaxial reactor, where they are heated in the presence of oxygen (O,1). The product of this oxidation is a metal oxide soot which is subsequently fused to form a high purity glass.

**VRAM** Video Random Access Memory. Similar in outward appearance to an ordinary 8-binary digit (BIT) random-access memory (RAM) a typical VRAM can be connected directly to the address and data bus of any bus-organized system, to provide composite video signals from microprocessor (MP) commands.

**VRC** (1) Vertical Redundancy Check. A parity check carried out on individual characters within a message block. It tests whether the sum of the binary digits (BITs) in a string of binary digits is odd or even.

(2) Voice Recognition Chip. When included with other components in a system, a typical voice recognition chip (VRC) can provide a vocabulary of 100 discrete words or phrases. The overall system formed is speaker dependent and is trained for the given vocabulary by the user.

**VS** Virtual System. A computer system that employs a method of operation such that it appears to the user to be accessing more main storage than the system actually contains. *See* VM, virtual memory.

**VSB** Vestigal Side Band. A system of amplitude modulation (AM) transmission in which one side band is transmitted in full, and the other side band (usually that corresponding to the lower modulating frequencies) is reduced in magnitude to a vestige. The attenuation is due to some of the spectural components of the side band, which are normally those relating to higher frequencies of the modulating wave. It is extensively used for television (TV) transmission.

**VSS** Voltage for Substrate and Sources. The common ground employed in metal-oxide semiconductors (MOS) circuits.

**VSYNC** Vertical SYNChronization. A system employed to ensure that the vertical movement of an electron beam in a cathode-ray tube (CRT) is locked to an applied signal. In television it is called field synchronizing and the specific signal is employed to determine the vertical position of an image displayed.

**VT** (1) Vertical Tabulation. A specific control character that is employed to shift a printing or display (D,6) position a predetermined amount perpendicular to the printing or display line.

(2) Voltage Transformer. An instrument transformer used for the transformation of voltage, designed and operated similarly to a power transformer. It is also known as a potential transformer. Often used in power or test circuits, the device reproduces the voltage (V,2) and current (I) values of a circuit.

**VTAM**  (1) Virtual Telecommunications Access Method. Teleprocessing input/output (I/O) software that uses virtual techniques (*see* VM). These techniques can be used to produce a desired end result, but the user need not know which is employed. They can make the memory larger than the real memory, the difference being made up by the software moving information rapidly in and out, to and from a backing store.

(2) Vortex TelecommunicAtions Method. A software package used in data communications that simplifies and organizes data communications programming to serve remote workstations for a host computer.

**VTOC**  Volume Table Of Contents. A list of the elements held on a computer disk directory.

**VTR**  (1) Video Tape Recorder. Reel-to-reel machines, i.e. equipment that uses open reels of tape, as opposed to those that use video cassettes or cartridges. (*See* VCR).

(2) Video Tape Recording. The action of using a video tape recorder (VTR,1).

**VU**  Volume Unit. A measure of audio signal levels. For a sinusoidal signal one volume unit equals one decibel (dB). VU meters are calibrated in volume units, which measure the total energy of an input signal rather than the peak and are extensively used in audio tape recording equipment. Also known as 'voice unit'.

**VXO**  Variable CRYSTAL Oscillator. A crystal controlled oscillator with an output than can be varied, but only over a limited range.

# W

**W**  (1) Tungsten. A heavy metal (atomic number 74) possessing an extremely high melting point and extensively used in the formation of filaments for bulbs. It is also widely used to form thermionic cathodes.

(2) Watt. The SI unit of power, it is a rate of energy transfer, of one joule per second. It can be defined as the rate of transfer or conversion of energy when a current (I) of one ampere (A) flows between two points which have a potential difference (PD) of one volt (V,1). The watt is also considered as the rate

of work, at one joule per second and therefore also has a mechanical definition.

**WACK** Wait before transmitting positive ACKnowledgement. Transmitted by a receiving station, this signal is used to indicate it is temporarily not capable of accepting a message.

**WADS** Wide Area Data Service. A data transmission service operating on a network similar to those serving a wide area telephone service (WATS) in the US.

**WAN** Wide Area Network. A computer network that covers a large physical area, even an entire continent. It has come into common use to make a distinction from local area network (LAN).

**WARC** World Administration Radio Conference. An international organization concerned mainly with the allocation of frequencies, but concentrating on those involved with satellite communications.

**WATFOR** WATerloo FORtran. Developed at the University of Waterloo in Canada, this is a version of formulae translator (FORTRAN) which provides a high rate of throughput.

**WATS** Wide Area Telephone Service. US telephone companies provide this service that enables customers to make calls to telephones in a specific zone on a dial basis for a flat monthly charge, using an access line. The WATS system divides the USA into six zones.

**Wb** WeBer. The SI unit of magnetic flux. The flux is measurable in terms of volt/seconds, and 1 Wb = 1 Vs.

**W band** Microwave ($\mu$W) frequencies occurring between 56 and 100 gigahertz and forming part of a larger continuous range of frequencies covering 0.225 to 100 gigahertz (GHz) which are usually subdivided into bands designated by letters. W band frequencies have wavelengths of between 0.536 and 0.3 cm.

**WCS** Writable Control Store. Enabling a user to alter dynamically a microprogram, this control store is implemented with a random-access memory (RAM).

**WDC** World Data Centre. One of a number of international stations for promotion and collection of scientific information.

**WDM** Wavelength Division Multiplexing. A method where more than one signal is transmitted on a channel (CHNL(a)) by dividing the available transmission frequency into narrower bands. A range of separate single channels are thus derived from splitting up a wider bandwidth into several narrower bandwidths. Also known as 'frequency-division multiplexing' (FDM).

**WEG** Wind Energy Group. A consortium set up by Taylor Woodrow, GEC and British Aerospace in the UK for the development of medium to large scale wind powered electricity generating systems.

**WIP** Work In Progress. Generally used to describe circuits (CCTs) or sub-assemblies in the production environment. It can refer to items passing through an item of automatic test equipment (ATE) or a piece travelling through a manufacturing process.

**WISE** World Information Systems Exchange. Arranged to encourage international collaboration on the exchange of data between a large number of institutions.

**WM** Word Mark. An indicator used to signal the beginning or end of a word.

**WOOFER** A low-frequency loudspeaker seldom used alone. This relatively large audio reproduction loudspeaker is designed to operate on relatively low frequencies in high-fidelity sound reproduction. Often it is part of a larger loudspeaker assembly comprising two or more speakers (*see* TWEETER).

**WP** Word Processing. Specific operating systems that include interactive information-retrieval systems, management information systems, typesetting, text editing and translation.

**WPI** World Patents Index. A store of information held on computer files containing patent information from all the major nations involved with research and development (R&D).

**WPM** Words Per Minute. A measure of speed in telegraph systems, describing the operating speed of teletypewriters (TTY) and computer printers.

**WRU** Who aRe yoU. Transmitted from a data terminal or telegraph, this signal is employed to initiate the automatic return transmission of an answer-back code from the called terminal. As this control character requests a telecommunication device to identify itself, it provides a level of security.

**WS** Working Storage. An area on a storage disk employed to hold dynamic or working data. It is contrasted with the reserved area that contains permanent information and the user area that contains semipermanent storage. It is also known as *working space*.

# X

**X band** Microwave ($\mu$W) frequencies occurring between 5.2 and 10.9 gigahertz (GHz) and forming part of a larger continuous range of frequencies covering 0.225 to 100 gigahertz which are usually subdivided into bands designated by letters. X-band frequencies have wavelengths between 5.77 and 2.75 cm.

**XEC** EXEcute register Contents. Allowing a programmer to

load a machine code into a working register, this instruction then permits the central processor unit (CPU) to treat the contents of the working register as an instruction.

**XFMR** TRANSForMeR. A static electromagnetic device consisting of two electric windings or circuits linked magnetically by an iron core. It uses an alternating voltage applied to one of the windings to produce, by electromagnetic induction, a corresponding voltage in the other winding.

**X-OFF** TRANSmitter OFF. (a) A signal that causes a transmitting circuit or unit to cease functioning. (b) An indicating device that signals that a transmitting device is no longer functioning.

**X-ON** TRANSmitter ON. (a) A signal that causes a transmitting circuit or unit to function. (b) An indicating device for signalling that a transmitting device is functioning.

**XOR** eXclusive OR. A logic operation where a true output (O,2) exists only when the input statements are different or odd.

**XPD** CROSSPolarization Discrimination. A measure of the purity of polarization of an antenna, usually expressed in decibels (dB). It is the direction that yields the maximum available power output (O,2) corresponding with the maximum received power, for that part of a radio system that is employed to emit radio waves and/or to extract energy from incoming radio waves.

**XPI** CROSSPolarization Interference. Co-channel interference that exists between a pair of orthogonally polarized signals in a radio system, that uses two signals having similar frequencies but opposite polarization.

**X-position** A punch position, also known as the 'eleven position', this is the second row of punch positions on a punch card.

**XPT** CROSSPoinT. An electronic or mechanical switching element that can be used to extend signal and speech paths of a connection.

**X-punch** An X-position hole; i.e. a hole punched in the X-position (*see* X-position) of a punched card.

**X-ray** An electromagnetic wave that has a wavelength between 0.01 and $100 \times 10^{-10}$ metres (m) with a frequency of 0.03 and $300 \times 10^{18}$ hertz (Hz). The energy can penetrate directly through substances that have small or negligible attenuation but not through high-density substances such as lead and gold.

**X series** Transmission recommendations. Concerned with information transmission over public data networks, this is a series of recommendations made by the Comité Consultatif Internationale de Télégraphique et Téléphonique (CCITT).

**X-Y plotter** Co-ordinates plotter. Used in conjunction with a computer, this device is able to plot co-ordinate points in the form of a graph.

**Y** (1) A star-connected three-phase power circuit.
(2) A transmitted signal that contains the brightness and detail information in colour television (TV) broadcasts.

**YIG** Yitrium-Iron-Garnet. A ferrite material widely used for microwave ($\mu$W) applications. It has magnetic properties that can be altered by the amount of trace elements it contains, the most common being calcium, vanadium and bismuth. The material is used as thin magnetic film on a nonmagnetic substrate in the production of solid-state magnetic circuits such as magnetic bubble memories.

**Y-position** A punch position, also known as the 'twelve position', this is the top row of punch positions on a punch card.

**Y-punch** A Y-position hole; i.e. a hole punched in the Y-position (*see* Y-position) of a punched card.

**Z-clipping** A computer-aided design (CAD) term. It describes the ability to specify depth parameters for a three-dimensional drawing such that all elements above or below this specified depth or depths become invisible. No alteration occurs to the database of the specific part or drawing.

**ZEF** Zero Extraction Force. Appertaining to connectors that require zero or minimum force to break apart mating items.

**ZIF** Zero Insertion Force. Appertaining to connections that require zero or a minimum of insertion force when being made. Typical units could be of the cam type of socket used for dual-in-line packages (DIP) or of the gas-tight high-pressure (GTH) design.

**ZIF/ZEF** Zero Insertion Force/Zero Extraction Force. Typically ZIF/ZEF connectors or sockets employ a cam so that mated connections are moved from their normal position, thus during mating or breaking apart there is no friction forces or springs to overcome. Once fully positioned, the cam is released allowing the contacts to mate. This type of connector has an inherent disadvantage in that its action does not provide a wiping action which in other connectors (*see* GTH) helps to clean the contacting or mating areas. However, this also means that there is very little if any contact wear. As a high number of cycles can be performed by ZIF/ZEF connectors, they are finding great popularity for applications in automatic test equipment (ATE).

**ZIL** Zigzag-In-Line. A package designed to achieve high-density installation at lower prices than existing dual-in-line (DIL) package chips. It is achieved by bending alternate legs in opposite directions, i.e. staggering them.

**Zn**  ZiNc. A metal with atomic number 30 used as an electrode in some electrolytic cells.

**Zr**  ZiRconium. A metal with atomic number 40 that can be employed as a material because it has strong chemical affinity for other materials (a getter), in hard vacuum electron tubes.

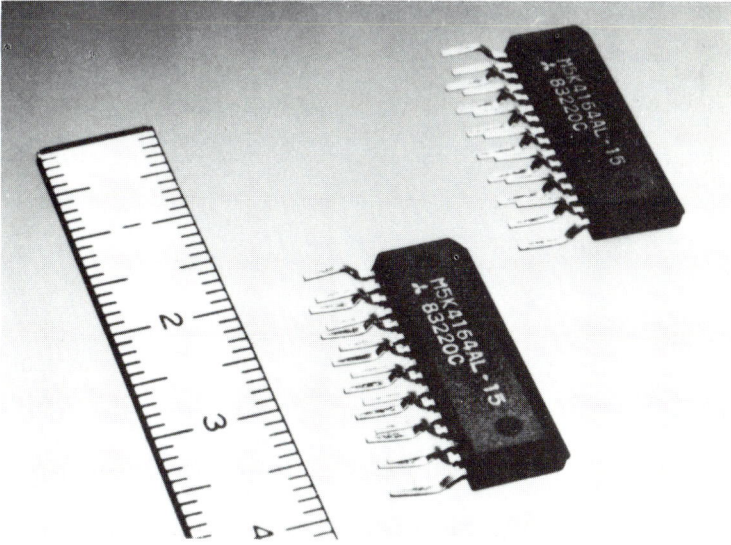

*Figure Z.1*  An example of a ZIL package showing the staggered legs (*courtesy* Mitsubishi Electric)

# Appendix

The British Standard for Primary Batteries, BS 397:1983, Part 1, 9.9.4, para. 3, states:

> The battery manufacturer should provide sufficient information to enable the retailer to select the correct battery for the user's application. This is specially important when supplying the first batteries for newly-purchased equipment.

Such a principle is fine – if it were adhered to – but the plethora of battery manufacturers and their products can be confusing. To help unravel any confusion, this Appendix provides a comparison of battery designations for the most popular types of battery supplied by the major producers.

In addition, the following advice concerning the care and use of batteries should help increase their life and protect equipment and its users.

Before batteries are inserted, the contacts of both equipment and batteries should be checked for cleanliness. If necessary, the contacts should be rubbed with either a clean pencil-eraser or a rough cloth, or possibly cleaned with a damp cloth and dried before being inserted.

It is extremely important that batteries be inserted correctly with regard to polarity (+ and −). One of the most common causes of dissatisfaction after replacement of a set of batteries is the reversed placement of one or more of them. In general, battery compartments are clearly and permanently marked to show the correct orientation of the batteries, and some are designed so that the batteries cannot be incorrectly orientated. Failure to comply with the instructions which are usually supplied with equipment can result in malfunction of and damage to the equipment.

It is not good practice either to use or to leave equipment exposed to extreme conditions, e.g. immediately on radiators, in a car parked in the sun, etc.

It is advantageous to remove batteries immediately from equipment which has ceased to function satisfactorily, or when a long period of disuse is anticipated, e.g. cine-camera, photo-flash, etc. Batteries should be removed from equipment also when mains current or other external source is being used to power it.

Although most batteries now on the market are provided with

protective jackets or other means to contain leakage, a battery that has been partially or completely exhausted may be more prone to leak than is an unused one.

All batteries in a set should be replaced at the same time since newly-purchased and partially-exhausted units should not be mixed. Also, batteries of different electrochemical systems, grade and bands should not be mixed. Failure to observe these precautions may result in some batteries in a set being driven beyond their normal exhaustion point and thus increase the probability of leakage.

No attempt should be made (1) to revive exhausted batteries by heat or other means, (2) to charge primary batteries as to do so may cause leakage and/or explosion, (3) to recharge batteries unless they are marked 'reachargeable'. Incorrect recharging may cause rupture.

Test meters do not provide reliable comparison of service to be expected from good batteries of different grades and manufacture. They do, however, detect serious failure.

The storage area for batteries should be clean, cool, dry, ventilated and weatherproof. Batteries should not be allowed to become short-circuited, e.g. by storing them in direct contact with metal objects.

Finally, batteries must never be disposed of in a fire – they may EXPLODE.

| Manufacturer | Sizes | | | | |
|---|---|---|---|---|---|
| Berec Ever Ready | HP7 | HP16 | SP11 HP11 | SP2 HP2 | PP3 |
| Daimon | 242 | 243 | 241 | 240 | 244 |
| Duracell Hellesens | MN1500 | MN2400 | MN1400 | MN1300 | MN1604 |
| Hi-Watt IEC | LR6 | LR03 | LR14 | LR20 | 6LR61 6LF22 |
| Maxell | SUM3 | SUM4 | SUM2 | SUM1 | 006P |
| Mazda | LK6 | LK03 | LK14 | LK20 | LK622 |
| Panasonic | AM3 | AM4 | AM2 | AM1 | 6AM6 |
| Philips | LR6 | LR03 | LR14 | LR20 | 6F22S |
| Ray-O-Vac | 815 | 824 | 814 | 813 | AC3 |
| Saft Leclanche | K6 | K03 | K14 | K20 | K622 |
| Toshiba | AM3 | AM4 | AM2 | AM1 | AM6 |
| Ucar | E91 | E92 | E93 | E95 | 522 |
| Usa | AA | AAA | C | D | |
| Varta | 4006 | 4003 | 4014 | 4020 | 4022 |
| Vidor | HP7 | HP16 | SP11 HP11 | SP2 HP2 | PP3 |
| Wonder | KLR6 | KLR03 | KLR14 | KLR20 | KLR22 |

## Specifications

| | Battery size | | | | |
|---|---|---|---|---|---|
| | AA | AAA | C | D | PP3 |
| Nominal voltage | 1.5 | 1.5 | 1.5 | 1.5 | 9.0 |
| Volume (cm$^3$) | 7.5 | 3.5 | 26.0 | 52.0 | 21.0 |
| Weight (grams) | 23.0 | 12.0 | 66.0 | 132.0 | 45.0 |
| IEC Designation | LR6 | LRO3 | LR14 | LR20 | 6F22 |

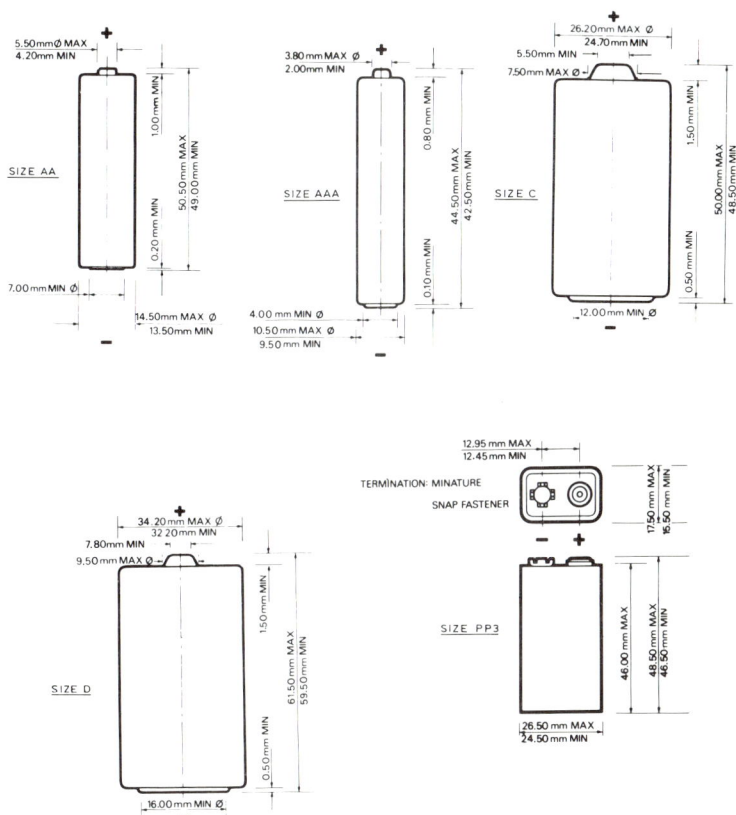

SIZE AA

5.50mm Ø MAX
4.20mm MIN

1.00 mm MIN

50.50mm MAX
49.00mm MIN

0.20 mm MIN

7.00 mm MIN Ø

14.50mm MAX Ø
13.50mm MIN

SIZE AAA

3.80 mm MAX Ø
2.00mm MIN

0.80 mm MIN

44.50mm MAX
42.50mm MIN

0.10 mm MIN

4.00 mm MIN Ø
10.50 mm MAX Ø
9.50 mm MIN

SIZE C

26.20mm MAX Ø
24.70mm MIN

5.50mm MIN
7.50mm MAX Ø

1.50 mm MIN

50.00mm MAX
48.50mm MIN

0.50 mm MIN

12.00 mm MIN Ø

SIZE D

34.20mm MAX Ø
32.20mm MIN

7.80mm MIN

9.50 mm MAX Ø

1.50 mm MIN

61.50mm MAX
59.50mm MIN

0.50 mm MIN

16.00 mm MIN Ø

TERMINATION: MINIATURE
SNAP FASTENER

12.95 mm MAX
12.45 mm MIN

17.50 mm MAX
16.50 mm MIN

SIZE PP3

46.00 mm MAX
48.50 mm MAX
46.50 mm MIN

26.50 mm MAX
24.50 mm MIN

237